Governing Military Technologies in the 21st Century

Other Palgrave Pivot titles

Thomas Birtchnell and William Hoyle: *3D Printing for Development in the Global South: The 3D4D Challenge*

David Fitzgerald and David Ryan: *Obama, US Foreign Policy and the Dilemmas of Intervention*

Lars Elleström: *Media Transformation: The Transfer of Media Characteristics Among Media*

Claudio Povolo: *The Novelist and the Archivist: Fiction and History in Alessandro Manzoni's The Betrothed*

Gerbrand Tholen: *The Changing Nature of the Graduate Labour Market: Media, Policy and Political Discourses in the UK*

Aaron Stoller: *Knowing and Learning as Creative Action: A Reexamination of the Epistemological Foundations of Education*

Carl Packman: *Payday Lending: Global Growth of the High-Cost Credit Market*

Lisa Lau and Om Prakash Dwivedi: *Re-Orientalism and Indian Writing in English*

Chapman Rackaway: *Communicating Politics Online*

G. Douglas Atkins: *T.S. Eliot's Christmas Poems: An Essay in Writing-as-Reading and Other "Impossible Unions"*

Marsha Berry and Mark Schleser: *Mobile Media Making in an Age of Smartphones*

Isabel Harbaugh: *Smallholders and the Non-Farm Transition in Latin America*

Daniel A. Wagner (editor): *Learning and Education in Developing Countries: Research and Policy for the Post-2015 UN Development Goals*

Murat Ustaoğlu and Ahmet İncekara: *Islamic Finance Alternatives for Emerging Economies: Empirical Evidence from Turkey.*

Laurent Bibard: *Sexuality and Globalization: An Introduction to a Phenomenology of Sexualities*

Thorsten Botz-Bornstein and Noreen Abdullah-Khan: *The Veil in Kuwait: Gender, Fashion, Identity*

Vasilis Kostakis and Michel Bauwens: *Network Society and Future Scenarios for a Collaborative Economy*

Tom Watson (editor): *Eastern European Perspectives on the Development of Public Relations: Other Voices*

Erik Paul: *Australia as US Client State: The Geopolitics of De-Democratization and Insecurity*

Floyd Weatherspoon: *African-American Males and the U.S. Justice System of Marginalization: A National Tragedy*

Mark Axelrod: *No Symbols Where None Intended: Literary Essays from Laclos to Beckett*

Paul M. W. Hackett: *Facet Theory and the Mapping Sentence: Evolving Philosophy, Use and Application*

Irwin Wall: *France Votes: The Election of François Hollande* palgrave▸pivot

Governing Military Technologies in the 21st Century: Ethics and Operations

Richard Michael O'Meara
Visiting Faculty, Rutgers University, USA

palgrave
macmillan

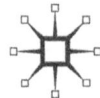

GOVERNING MILITARY TECHNOLOGIES IN THE 21ST CENTURY
© Richard Michael O'Meara, 2014.

All rights reserved.
First published in 2014 by
PALGRAVE MACMILLAN®
in the United States—a division of St. Martin's Press LLC,
175 Fifth Avenue, New York, NY 10010.

Where this book is distributed in the UK, Europe and the rest of the world, this is by Palgrave Macmillan, a division of Macmillan Publishers Limited, registered in England, company number 785998, of Houndmills, Basingstoke, Hampshire RG21 6XS.

Palgrave Macmillan is the global academic imprint of the above companies and has companies and representatives throughout the world.

Palgrave® and Macmillan® are registered trademarks in the United States, the United Kingdom, Europe and other countries.

ISBN: 978-1-137-44918-4 EPUB
ISBN: 978-1-137-44917-7 PDF
ISBN: 978-1-137-44916-0 Hardback

Library of Congress Cataloging-in-Publication Data is available from the Library of Congress.

A catalogue record of the book is available from the British Library.

First edition: 2014

www.palgrave.com/pivot

DOI: 10.1057/9781137449177

> The human race has reached a turning point. Man has opened the secrets of nature and mastered new powers. If he uses them wisely, he can reach new heights of civilization. If he uses them foolishly, they may destroy him. Man must create the moral and legal framework for the world which will insure that his new powers are used for good and not for evil.
>
> Harry S. Truman

Contents

Acknowledgments		vii
About the Author		viii
List of Abbreviations		ix
1	Introduction: The Nature of the Problem	1
2	Gadgets and Gizmos	10
3	Innovators and Consumers: The Culture of Innovation and Use of Military Technology in the 21st Century	24
4	Intended and Unanticipated Consequences	44
5	Contemporary Governance and Architecture	66
6	Arms around the Problem: Suggestions for Future Governance	80
7	Conclusion	99
Bibliography		104
Index		121

Acknowledgments

To my mentors, Simon Reich, Ed Barrett, Mike Romeo, Richard Langhorne, Yale Ferguson, and Kerryann O'Meara who have made an academic out of an old soldier; to the soldiers with whom I have served, especially in the First Infantry Division and the 4th MLC; and to Mary Jean whose constant support makes all things possible.

About the Author

Richard M. O'Meara is a retired US Army Brigadier General and trial attorney who teaches global and regional studies at Rutgers University and Richard Stockton College of New Jersey. He has presented and published widely on the ethical issues which are involved in the application and use of military technologies.

List of Abbreviations

AAA	American Anthropologists Association
AI	artificial intelligence
AW	autonomous weapon
BWC	Convention on the Prohibition of the Development, Production and Stockpiling of Bacteriological (Biological) and Toxin Weapons and on Their Destruction
CBNR	chemical, biological, radiological and nuclear reconnaissance
CCW	Convention on Prohibitions or Restrictions on the Use of Certain Conventional Weapons Which May Be Deemed to Be Excessively Injurious or to Have Indiscriminate Effects
CETMONS	Consortium for Emergency Technologies, Military Operations and National Security
COGOMS	combatant commanders
CTBT	Comprehensive Test Ban Treaty
CWC	Convention on the Prohibition of the Development, Production, Stockpiling and Use of Chemical Weapons and on Their Destruction
DARPA	Defense Advanced Projects Agency
DOD	(US) Department of Defense
EMP	electromagnetic pulse
EOD	explosive ordnance disposal
GPS	global positioning system
GWOT	global war on terrorism
HTS	human terrain system program

ICBM	intercontinental ballistic missile
ICC	International Criminal Court
ICJ	International Court of Justice
IED	improvised explosive devise
IHL	International Humanitarian Law
IT	information technology
KMD	knowledge-enabled mass destruction
LAR	lethal autonomous robot
LOAC	law of armed conflict
MNT	molecular nanotechnology
NBIC	nano, bio, info, and cognitive technologies
NCA	network of concerned anthropologists
NCW	network-centric warfare
NPT	Treaty on the Non-Proliferation of Nuclear Weapons
NT	nanotechnology
PGM	precision-guided missiles
PTSD	post traumatic stress disorder
R&D	research and development
RMA	Revolution in Military Affairs
SIGNIT	signals intelligence
UAS	unmanned aerial system
UGV	unmanned ground vehicle
UMS	unmanned maritime system
USAF	United States Air Force
WMDs	weapons of mass destruction

1
Introduction: The Nature of the Problem

Abstract: *The innovation, adaption, and use of new technologies on the battlefield have a history which precedes even the written record. How to identify the nature of their impact and regulate their use has been an ongoing challenge for centuries. Contemporary innovation, however, may well represent a shift in traditional paradigms, given that innovation is democratized, that is available to anyone with minimal constraints, and flourishes in unregulated spaces. How to describe emerging technologies and their impact on the battlefield is of considerable importance as humankind wrestles with the legal and ethical decisions required to insure that it is not overtaken by, and perhaps destroyed by, technology's unintended potential.*

O'Meara, Richard Michael. *Governing Military Technologies in the 21st Century: Ethics and Operations.* New York: Palgrave Macmillan, 2014.
DOI: 10.1057/9781137449177.0005.

The scope of contemporary technological innovation is both impressive and staggering. Indeed, for the average consumer of these technologies, whether on the battlefield or in daily life—the general who orders this technology, the politician who pays for it, the user whose life is changed by it, even the Luddite who rails against it—these technologies are magic. They are incomprehensible in the manner of their creation, the details of their inner workings, the sheer minutiae of their possibilities. They are like the *genie* out of the bottle clamoring to fulfill three wishes; guess right and the world is at your fingertips, guess wrong and there may well be catastrophe. And you have to guess quickly for the *genie* is busy and has to move on. There are, of course, shamans who know the *genie's* rules, who created the *genie* or at least discovered how to get it out of the bottle. You go to them and beg for advice regarding your wishes. What should I take from the *genie*? How should I use my wishes? Quickly tell me before I lose my chance and the *genie* makes the choices for me. And you find that the shaman is busy with new *genies* and new bottles and hasn't given your choices much thought at all. He may stop to help you ponder your questions, but most probably he goes back into his tent and continues his work. "You're on your own kid... Don't screw up!"

Discussions regarding the scope of emerging technologies are often difficult due to the breadth and sophistication of the information about them. They often descend into ramblings about gadgets and gizmos and reflect the short answer to Peter Singer's question, "Why spend four years researching and writing a book on new technologies? Because robots are freakin' cool."[1] Because innovation is and has always been catalytic, feeding off itself, reacting to its intended and unintended consequences, influenced by the environment in which it is created and creating new environments as it goes, the discussion must, or course, be much longer and more nuanced. Of equal importance is the fact that demands for emerging technologies are coming faster and faster, and failure to keep up can have disastrous effects on the battlefield.

Emerging technologies are not new to the battlefield. Indeed, humankind has effected and been affected by the military use of technology since at least the use of stones as weapons. The history of warfare, of course, begins with the written record. The anthropology of warfare, on the other hand, has a considerably longer tail. While there is a good deal of speculation regarding man's inherent penchant for violence, it is clear that many of the characteristics of the successful warrior can also be identified with the successful hunter of the prehistoric age. Prehistorians

Henri Breuil and Raymond Lautier, for example, note the similarities between hunting humans and their prey. "The bonds between them were not yet broken, and man still felt near to the beasts that lived around him, that killed, and fed him... from them he still retained all the faculties that civilization has blunted—rapid action and highly trained senses of sight, hearing and smell, physical toughness in an extreme degree, a detailed, precise knowledge of the qualities and habits of game, and great skill in using with the greatest effect the rudimentary weapons available."[2]

Yet, there are many ways to look at the issue. John Keegan notes that some 10,000 years ago there occurred, perhaps for the first time, a revolution in weapons technology with the appearance of four "staggeringly powerful new weapons"—the bow, the sling, the dagger, and the mace.[3] And there have been multiple revolutions since. Keegan, for example, divides his discussion of warfare into four general groups: stone, flesh, iron, and fire. These categories refer to the types of technologies described and their impact on civilization. *Flesh,* for example, speaks to the harnessing of animals, specifically horses and the technologies used by horse warriors; chariots, warhorses, large and small, and the composite bow (used on horseback by nomads), and so on.[4] Some innovations, according to Keegan, are so revolutionary as to change the manner in which mankind operates. Geography, too, affected the tools man developed in order to challenge those privileged to live in river valleys where stone, bronze, and the horse monopolized the successful organization of war. Iron, it turns out, was a game changer in this regard. Unlike bronze, a difficult technology both in its forging and its cost, iron appears to have been much more available. Because it created stronger tools and was able to maintain a sharper edge, its use quickly became widespread and changed multiple social relationships both between the outliers of the river civilizations and within outlier societies. The adaption of iron to agriculture, for example, opened up the vast heartland as iron tools were used to tackle the difficult soils of the steppe. Iron swords, shields, helmets, and spears became the tool of choice for soldier formations, which because of these tools' relative cost, became larger as more and more men could afford their ownership and use.[5]

This discussion, then, emphasizes the technology itself and chronicles the myriad intended and unanticipated consequences that flow from its creation and use.

More recently, the literature eschews discussions of specific technologies and speaks in terms of *industrial revolutions* and their effect on tactics and strategy with the warning that actors (traditionally nation-states but increasingly non-state actors) which are unable to recognize the importance of technology and adapt accordingly "... cease to be great...".[6] "Great powers," Max Boot argues, "cease to be great for many reasons. In addition to the causes frequently debated—economics, culture, disease, geography—there is an overarching trend. Over the last 500 years, the fate of nations has been increasingly tied to their success, or lack thereof, in harnessing revolutions in military affairs."[7] Here the emphasis is not on the particular technology itself, but rather the ability of the group to envision and organize its application, conceive of its relationship and use with other technologies, and otherwise maximize its benefits as it competes with other groups.

Others continue the de-emphasis of specific technologies and speak of *military-social revolutions*. Williamson Murray, for example, emphasizes the creation of the modern state and its ability to organize both warfare and peaceful endeavors. "Between 1792 and 1815, two separate military-social revolutions occurred which again altered the framework of war. The French Revolution completely upset the social and political framework within which the European states had conducted their wars since the Treaty of Westphalia in 1648 [institution of the practice of *levée en masse* and total war], and the Industrial Revolution was to have equally profound implications...

> If technology exercised little influence over the battlefields of this period, it did play a crucial part in the Seventh Coalition's winning the campaign against Napoleon. The Industrial Revolution was at the time changing the way the British economy worked. By revolutionizing the means of production, it altered the basis on which economic activity had rested since the dawn of time—namely, human, and animal muscle power. The gains this revolution in economic affairs and technology provided to Britain enabled its government to subsidize the great coalitions against the French, including the last one that destroyed Napoleon's empire.[8]

Murray continues his emphasis regarding the symbiotic relationship between the innovation of emerging technologies in the civilian sphere and in the military. "From 1914 to 1989, military technology drove civilian technology. During the interwar years, military organizations pushed the development of technologies like the airplane and radio, with spin-offs like radar, all of which had immense significance for civilians... Today, the 1914–89 pattern has shifted back to the pre-1914 paradigm: technological developments in the civilian world of computers and communications are now driving military technology."[9] Another way of getting at the subject is to speak the language of epidemiology and ecology. Here mankind constitutes the only significant macroparasites of the animal world

> ... who, by specializing in violence, are able to secure a living without themselves producing the food and other commodities they consume. Hence a study of macroparasitism among human populations turns into a study of the organization of armed force with special attention to changes in the kinds of equipment warriors used. Alterations in armaments resemble genetic mutations of micro-organisms in the sense that they may, from time to time, open new geographic zones for exploration, or break down older limits upon the exercise of force within the host society itself.[10]

Others speak of *revolutions in military affairs* or cultural *ways of war*. Peter Wilson, for example, outlines four revolutions in military affairs which have resulted in four ways of war for the United States. The first deals with organization around fighting vehicles and the way they communicate; the second deals with irregular warfare (which he believes will become the *regular* way of war); the third involves the standoff in nuclear weapons technology, practiced primarily during the Cold War; and the fourth deals with the present and includes military operations at high speed with low casualties which are rapid and decisive.[11] All of these involve responses to new military technologies and demonstrate, in his view, attempts to harness and organize around these technologies in order to gain efficiencies on the battlefield. Michael Guetlein speaks specifically to information technology when he notes, "[T]he business of collecting, communicating, and processing information will become its own dimension of warfare. Information systems combined with rapid decision support tools integrated onto a single platform are already driving a revolution in military affairs (RMA)."[12] "Presently," according to another commentator, "RMA technologies are changing the nature of war-waging by enabling precise destruction of targets from a distance and speeding up the processes of decision making. The quest for modernization caters for (sic) emerging capabilities of states' potential adversaries, cost factors, and raising the technological threshold of armed forces. This advent of the RMA clearly indicates how technology plays an important role in regard to national security."[13] Gotz Neuneck and Christian Alwardt sum up the discussion regarding RMA as they emphasize three distinct changes, administrative, technological, and ideological. Examples include the French Revolution's *levée en mass* which was able for the first time to mobilize the entire resources of a state and dedicate them toward the conduct of war. Technologies such as telegraphy, mechanization of vehicles, underwater ships and planes, not to mention munitions including nuclear weapons have radically changed the manner in which states are able to project violence. And the development of ideologies, in conjunction with the ability of leaders to communicate on a mass scale, give meaning, direction, and purpose to individuals as they suffer the vagaries of total war. There is a distinction to be made between military revolutions and revolutions in military affairs. "Military revolutions such as the 'French Revolution' or the 'advent of nuclear weapons' are cumulative and hard to predicts in their consequences for modern states and societies. Revolutions in Military Affairs, on the other hand, reset in the defeat of enemies (e.g. the 1991 Iraq War), do not necessarily shape the character of states and societies."[14]

Whatever the analysis, it can be argued with a fair degree of certainty that the innovation of emerging technologies has been and will continue to be pervasive, and that their use has considerable impact on the ways humankind operates. Their use on the battlefield, moreover, often defines the ability of human organizations to survive and prosper.

Of considerable interest and concern is the increasing availability of emerging military technologies in the 21st century and their governance. Some have taken the position that political control has wrested from the innovators the ability to develop and proliferate new and damaging technologies. William McNeil, for example, in 1982, was able to conclude that once the state had mastered the ability to *control* technological change and bend it to its needs. "[H]uman society, in short, returned to normal. Social change reverted to the leisurely pace of preindustrial, precommercial times. Adaption between means and ends, between human activity and the natural environment, and among interacting human groups achieved such precision that further changes became both unnecessary and undesirable. Besides, they were not allowed."[15]

Others are not so sure. One neuroscientist worries about the projection of his science into the battlefield without discrimination. There continues as part of the definition of humanness to be a propensity for both the destructive and the constructive. "The long-term trajectory of humanity combines a growing capacity for indiscriminate destruction along with vast increase in constructive methods and techniques for solving problems that inhibit human flourishing. Somehow, these seemingly contradictory traits must be neurologically linked. Perhaps, understanding more about this excruciatingly complex system, we can turn ourselves from the wars of the mind to the peace of the soul."[16]

Regarding nanotechnology, another commentator cautions that governance is not only appropriate but necessary; lest we destroy those fundamental values, institutions, and habits of conduct which permit humankind to operate with relative freedom and success. "Through a variety of plausible mechanisms including pharmaceuticals, nano-enabled neural implants, and brain stimulation, the NBIC enhancement of human beings may allow for the greater exercise of human freedoms, but it holds potential for undermining liberal democratic values as well. In this fundamental ambiguity, such technologies require a significant degree of scrutiny—part of a process we will call 'anticipatory governance.'"[17]

There are those in the robotics community who sense that these emerging technologies may foreshadow something different than the traditional intersection between technology and the way humankind operates, both on the battlefield and in society generally. The stakes, according to this argument, are extremely high and, in the language of history, *axial*. Peter Singer, for example, notes:

> [H]umans have long been distinguished from other animals by our ability to create. Our distant ancestors learned how to tame the wild, reach the top of the food chain, and build civilization. Our more recent forebears figured out how to crack the codes of science, and even escape the bonds of gravity, taking our species beyond our home planet. Through our art, literature, poetry, music, architecture, and culture, we have fashioned awe-inspiring ways to express ourselves and our love of one another.
>
> And now we are creating something exciting and new, a technology that might just transform human's role in their world, perhaps even create a *new species* [emphasis added]. But this revolution is mainly driven by our inability to move beyond the conflicts that have shaped human history from the very start. Sadly, our machines may not be the only thing wired for war.[18]

And, finally, there are those who believe they see clearly to the bottom of the abyss and find no solace in the idea that humankind has always found a way to master emerging technologies. Bill Joy, for example, a self-styled *generalist*, warns of a different set of threats than mankind has ever experienced. We are living in a world where scientific breakthroughs are ubiquitous and, for this reason, they appear unthreatening or at least their rapid creation and use often goes unnoticed. There is, according to Joy, a difference between previous eras of technological advancement and the present. Engineered organisms such as robots and nanorobots are both intelligent and self-replicating. As these technologies gain independence from their creators through small, seemingly sensible improvements, there is the potential, indeed the inevitability, of great danger. They are available, for example, to multiple users who may or may not be interested in good governance. Knowledge alone appears to be the ticket to their use.

> Thus, we have the possibility not just of WMDs but of knowledge-enabled mass destruction (KMD), this destructiveness hugely amplified by the power of self-replication.
>
> I think it is no exaggeration to say we are on the cusp of the further perfection of extreme evil, an evil whose possibility spreads well beyond that which WMDs bequeathed to the nation-states, on to a surprising and terrible empowerment of extreme individuals.[19]

If innovation of these emerging technologies is indeed *democratized*, that is available to anyone with minimal constraints; if innovation is best encouraged in *fragmented* and competitive environments; and finally, if innovation flourishes best in unregulated spaces, the room for the creation of Joy's *extreme evil* would appear to be great, with no hope of putting the *genie* back in the bottle. Finding the balance between the freedom to *innovate* and the identification of places where innovation *should not go* would appear to be not only rational but necessary.

One approach is to look at four separate areas of the problem: first, the types of technologies which are making their way onto the battlefield; second, the environment in which they are created and used; third, the intended and unanticipated consequences of emerging military technologies; and fourth, the tools available to regulate their innovation and use.

Notes

1   P. W. Singer *Wired for War: The Robotics Revolution and Conflict in the 21st Century* (New York: Penguin Group, 2009), 1.
2   Henri Breuil and Raymond Lautier, trans. B.B. Rafter. *The Men of the Old Stone Age* (Westport, CN: Greenwood Press, 1980), 71.

3   John Keegan. *A History of Warfare* (New York: Knopf, 1993), 118.
4   Ibid, 155.
5   Ibid, 237.
6   Max Boot. "Are We the Mongols of the Information Age?" *Los Angeles Times*, op-ed (October 29, 2006), retrieved at http://www.cfr.org/publicatins/11837/are_are_we_the_mongols_of_the_information_age.html, October 10, 2010.
7   Ibid., 1. See also Max Boot. *War Made New: Weapons, Warriors and the Making of the Modern World* (New York: Penguin, 2006).
8   Williamson Murray. "War and the West," *Orbis*, Philadelphia, Pa, no. 2 (Spring 2008), 350.
9   Ibid., 356.
10 William H. McNeill. *The Pursuit of Power, Technologies, Armed Force, and Society since A.D. 1000* (Chicago: The University of Chicago Press, 1982), vii.
11 Peter Wilson. "Revolutions in Military Affairs as Ways of War, 1914–2014." Presentation at Strategic Implications of Emerging Technologies Conference, XX Strategy Conference, U.S. Army War College, Carlisle Barracks, Pa (April 2009).
12 Mike Guetlein. "Lethal Autonomous Weapons—Ethical and Doctrinal Implications," Naval War College Joint Operations Paper, (February 2005), 6.
13 Ajey Lele. "Technologies and National Security," *Indian Defence Review*, Vol. 24.1 (Jan–Mar, 2009), 6.
14 Gotz Neuneck and Christian Alwardt. "The Revolution in Military Affairs, Its Driving Forces, Elements and Complexity." *Interdisciplinary Research Group: On Disarmament, Arms Control and Risk Technologies,* Working Paper #13 (May 2008).
15 McNeil. *The Pursuit of Power,* 386.
16 Jonathan D. Moreno. *Mind Wars, Brain Research and National Defense* (New York: Dana Press, 2006), 184.
17 David H. Guston, John Parsi, Justin Tosi. "Anticipating the Ethical and Political Challenges of Human Nanotechnologies," in Fritz Allhoff, Patrick Lin, James Moor, John Weskert. *Nanoethics, the Ethical and Social Implications of Nanotechnology* (Hoboken, NJ: John Wiley & Sons Inc., 2007), 185.
18 Singer. *Wired For War*, 436.
19 Bill Joy. "Why the Future Doesn't Need Us" in Allhoff, et al. eds. *Nanoethics*, 21–22. See also, Stephen Hawkings. "Transcendence looks at the implications of artificial intelligence—but are we taking AI seriously enough?" *The Independent* (May 4, 2014), retrieved at http://www.independent.couk/news/science/stephen-hawking-transcendence-looks-at-theimplications-of-artificial-intelligence, May 4, 2014.

2
Gadgets and Gizmos

Abstract: *Emerging technologies are often categorized in terms used by the specialists who engineer them. For the layman, however, a better approach is to speak in terms of nanotechnology, human enhancement technologies (including neuroscience), robotics, non-lethal weapons, and cyber technology. Each has applications which are presently on the battlefield and each carries with it its own governance concerns.*

> Part one: Nanotechnology
> Part two: Human Enhancement
> Part three: Robotics
> Part four: Non-lethal Weapons
> Part five: Cyber Technology

O'Meara, Richard Michael. *Governing Military Technologies in the 21st Century.* New York: Palgrave Macmillan, 2014. DOI: 10.1057/9781137449177.0006.

Emerging technologies are often categorized in terms of the specialists who engineer them; thus, we speak in terms of nano, bio, info, and cognitive technologies (NBIC). In terms of the most immediate effect on the battlefield, however, it is, perhaps, more appropriate to look at nanotechnology, human enhancement technologies (including neuroscience), robotics, non-lethal weapons, and cyber technology.[1] These are, in multiple forms, already on the battlefield; their uses are proven, and their possibilities identified for further use. Further, they are shaping and being shaped by the environments in which they have been employed.

Part one: Nanoscience

Nanoscience, in a sense, is not a separate discipline at all, but is rather, in part, a way of reducing the size of things—biological, robotic, and informational things amongst others. A nanometer measures one one-billionth of a meter. In comparison, a human hair is approximately 50,000 nanometers across or to put it another way, a nanometer is as much smaller than a football as a football is smaller than the distance from the earth to the moon. As a result, almost all molecules lie within the nanoscale which permits design with incredible precision. All materials, then, can be created with amazing properties. Thus, nanomaterials are considerably stronger than anything that has been made in the past. Indeed they are the strongest material it will ever be possible to make.[2]

Nanoscience, then, is used to change and make more efficient a wide range of other engineered technologies from fiber optics, to optics, to fabrication, to biology, to robotics. In application, nanotechnologies reduce the weight of objects and the speed with which information is transmitted. They contribute to the enhancement of humans through prosthetic technologies, and reduce the cost of making things generally. As Mark and Daniel Ratner predict their benefits "... will shift paradigms in biomedicine (e.g. imaging, diagnosis, treatment, and prevention; energy [e.g. conversion and storage]; electronics [e.g. computing and displays]); manufacturing; environmental remediation; and many other categories of products and applications." Perhaps the most important aspect of nanoscience and nanotechnology is the ability to respond to what might be called grand challenges. These are major problems such as diagnosing particular forms of cancer, stopping corrosion on metal bridges, providing early warning of heart malfunctions, developing environmentally friendly and significant new energy sources, providing total assurance of food safety, producing reliable long term storage of information, and so on.[3]

The point here is that nanoscience permits the creation of multiple capabilities on the entire spectrum of human endeavor through multiple disciplines with foreseeable and unforeseeable consequences. One example, the Ratners provide, speaks to the freewheeling character of this work. Scientists utilizing both the sciences of biology and engineering are creating noses with nanosized sensors that are capable of "sniffing out" smells that are otherwise imperceptible to humans. Artificial skins result in hypersensitivity and artificial eyes using the examples provided by insects radically alter the ability of humans to see in both daylight and darkness.[4]

Another example speaks to the intersection of nanotechnology, bioscience, and robotics.

In unconventional terms, bionanobots might be designed that, when ingested from the air by humans, would assay DNA codes and self-destruct in those persons whose codes had been programmed. Nanobots could attack certain kinds of metals, lubricants, or rubber, destroying conventional weaponry by literally consuming it.[5]

Nanoscience, then, speaks to the ability, at least in part, to miniaturize to a size well below the width of a human hair all manner of mechanical devices, thereby rendering them capable of super efficiencies on the battlefield. Not only can it perform functions, it also has the capability to create out of molecular redesign new materials which are stronger, more pliant, weigh less, and are capable of accomplishing more than materials used today. Indeed, one commentator speculates regarding the ability of the science to create a "nanofactory" in the future. While there are many designs available, one might envision a desktop device that can produce virtually any object needed from tank parts to foodstuffs. The ability to produce, copy, and then transport "the sinews of war" without the need for large logistics tails impacts the challenges of time, space, and cost which have traditionally hampered the ability of the state to put an army in the field and keep it there for any length of time.[6]

Part two: Human enhancement

Human enhancement generally and neuroscience specifically have as their goal, at least in part, intervention into the human organism for the purpose of changing it. One director of the Defense Advanced Research Projects Agency (DARPA) in 2003 advised Congress that the goal is to exploit "...the life sciences to make the individual warfighter stronger, more alert, more endurant, and better able to heal."[7] All military technology, throughout history, has had the goal of making the warfighter, whether he/she be in a plane, on the battlefield or under the sea, more effective, that is more capable of accomplishing the tasks necessary to compete and win military engagements. Thus, improved weapons systems, communications systems, uniforms, logistical capabilities (cleaner water, hotter food, etc.) and even propaganda have been designed with the enemy and the environment in mind. The justification, "if it saves American lives, do it!" is of particular relevance here. In a sense, then, human enhancement is nothing new. Its contemporary and future applications, however, are of significant interest.

A recent report by JASON, the MITRE Group, regarding human enhancement for the U.S. Office of Defense Research and Engineering states that there have been considerable advances in medical intervention for stroke recovery, spinal cord repair, development of prosthetics and neural interfaces for tetraplegics. Treatment of post-traumatic stress disorder, depression, Alzheimer's disease, and Parkinson's disease have all been impacted by improvements in psychopharmaceuticals and brain stimulation treatment. The report concludes that there are essentially two areas where there have been substantial advances in human modification. One is *brain plasticity* which changes the function of the human brain either by training or the application of pharmaceuticals. A second involves the interface of the brain with computers, essentially hooking up the nervous system to external devices in order to enhance the ability of the brain to interact with the outside world without dependence on the natural senses (hearing, seeing, smelling, touching, etc.).[8]

A short list of contemporary and envisioned human enhancements makes the point. Controlling fatigue and the poor decisions made as a result of sleep deprivation through pharmaceuticals; creating superior physical and psychological performance by controlling energy metabolism on demand (example: creating continuous peak performance and cognitive function for three to five days, 24 hours per day, without the need for calories); improving cognitive capacity through gene and proteomic medicine, emplacement of brain prostheses and training; controlling emotions such as fear and guilt through pharmaceuticals and gene therapy; emplacement of mechanical sensors and processors into the human body; and erasing bad memories through pharmaceuticals and electrical or magnetic stimulation; these, and a wide range of other examples, speak to what might be called *game-changing* technologies.[9]

Given the fact that the military is by its nature authoritarian, that is decisions regarding use of bioenhancement technologies are made top down rather than bottom up; and further, the fact that the military tends to subordinate all decisions to its one overriding *raison d'être* "to fight and win America's wars," military organizations are susceptible to the multiple complications and benefits which bioenhancement brings. The President's Council on Bioethics warns that bioenhancement presents multiple challenges going forward especially in the area of unintended consequences. First, there is the problem of unforeseen consequences. The Council points out that biotechnology, like all technologies, is not created for a particular use. Rather, the goals it serves and the powers that are derived from its applications are defined by human users. A technology created to serve one use, therefore, inevitably morphs into multiple unintended uses not all of which can be said to be advantageous. A second issue involves the overall goal of biotechnology to improve the lot of humankind. A seminal question is: *what is it about humankind that needs improvement*? A further question, political in nature is: *who gets to decide*? A final issue involves the question of metrics. How can we judge if a particular biotechnological change has, indeed, worked for the benefit of individuals and society generally?[10] Like, for the military, is the efficient accomplishment of the mission the only requirement for the application of biotechnologies on soldiers?

All the general goals of bioscience, increasing knowledge about the brain and the biological sources of human behavior, neuropharmacology and the manipulation of emotions and behavior, the prolongation of life, and genetic engineering are proving extremely relevant to the military project. The requirement to remain competitive with potential adversaries on the battlefield ensures their continued rapid development as well.[11]

Part three: Robotics

Robotics enjoy pre-eminence in the discussion of military technologies, perhaps, because popular culture has served to inform the public of their possibilities and, further, it may be said that their applications are easier to comprehend. *The Terminator, Matrix, Ironman, Star Trek, I, Robot, and Transcendence* all employ robots and aspects of artificial intelligence as central characters/leitmotifs and extol their virtues in multiple ways. A recent movie, *The Hurt Locker,* chronicles the relationship between an Army explosives expert and his robot as he goes about the business of dismantling Improvised Explosive Devices (IEDs) in Iraq. Predator drones appear in the news daily as they go about the business of identifying and engaging Taliban and Al Qaida targets in Afghanistan, Pakistan, Yemen, and elsewhere. Indeed, robots have been the subject of science fiction literature for decades.[12] Robots are, in essence a type of machine, but one that embraces the "sense-think-act" paradigm. Clearly they are man-made devices, although in the future they will most assuredly be made by other robots as robotics dominates the world of manufacture. They are comprised of three components: *sensors* that monitor the environment and, like natural sensors in the human body, detect changes in it; *processors* (or artificial intelligence) that analyze the information received by the sensors and act on it, that is make decisions about what if any changes to the environment need to occur; and *effectors* which act on the environment based on the decisions made by the *processors*. When these three parts act together, they provide for the robot the functional equivalent of an organism. They are stand alone devices which are capable of acting without human supervision.[13]

Robots are deployed to perform a wide range of tasks on and off the battlefield and Congress has mandated that their use expand radically in the next decade. The Department of Defense reports:

> In today's military, unmanned systems are highly desired by combatant commanders (COGOMs) for their versatility and persistence. By performing tasks such as surveillance; signals intelligence (SIGNIT), precision target designation, mine detection; and chemical, biological, radiological, nuclear (CBRN) reconnaissance, unmanned systems have made key contributions to the Global War on Terror (GWOT). As of October 2008, coalition unmanned aircraft systems (UAS) (exclusive of hand-launched systems) have flown almost 500,000 flight hours in support of Operations Enduring Freedom and Iraqi Freedom, unmanned ground vehicles (UGVs) have conducted over 30,000 missions, detecting and/or neutralizing over 15,000 improvised explosive devices (IEDs), and unmanned maritime systems (UMSs) have provided security to ports.[14]

It has been a longstanding goal of Congress to increase the use of robots in the military for some time. The Defense Authorization Act of 2000, for example, states that "[I]t shall be the goal of the Armed Forces to achieve the fielding of unmanned, remotely controlled technology such that—(1) by 2010, one-third of the aircraft in the operational deep strike force aircraft fleet are unmanned; and (2) by 2015, one-third of the operational ground combat vehicles are unmanned."[15]

Further, their development has increased as the needs have been identified. The Department of Defense reports that its investment in the technology has seen "... unmanned systems transformed from being primarily remote-operated, single-mission platforms into increasingly autonomous, multi-purpose systems. The fielding of increasingly sophisticated reconnaissance, targeting, and weapons delivery technology has not only allowed unmanned systems to participate in shortening the 'sensor to shooter' kill chain, but it has also allowed them to complete the chain by delivering precision weapons on target."[16] In other words, semi-autonomous robots are being used to kill enemies on the battlefield, based on information received by their sensors and decisions made in their processors.

Robots have multiple benefits. For one thing, they permit militaries to operate with fewer soldiers. As manpower pools for military recruitment shrink, it is expedient to substitute machines for soldiers in order to maintain military advantage. Second, robots are politically convenient. The 21st century, especially in liberal democracies like the United States, exhibits a distaste for large-standing armies and causalities. Robots, like private contractors, are not counted in national casualty reports nor are their wounds the subject of debate or scrutiny. Third, robots cost a good deal less than human combatants. Armin Krishner reports that the average soldier costs the nation approximately $4 million over his lifetime while the average robot might cost 10 percent of that figure.[17] In many ways they are simply more efficient than humans. Their sensors, for example, can gather infinitely more information than humans; their processors can make sense of that information by tapping into multiple information streams and databanks at a faster rate than humans; and their effectors can unleash appropriate responses to that information more efficiently than humans. Further, they don't carry with them the baggage of human frailty. They don't get hungry, they are capable of serving in adverse environments for extremely long periods of time, there is minimal question of dehydration, gender and other cultural issues, and they are not affected by the loss of soldiers around them. Perhaps more important, they will be able to self-replicate and maintain themselves.

In the future, robotists tell us that it is probable that robots, with the addition of artificial intelligence (AI), will be capable of acting independently, that is, without human supervision—called *humans in the loop*—in the accomplishment of most tasks presently performed by soldiers. One definition defines AI as "the science of making machines do things that would require intelligence if done by men."[18] AI, although not available today except in the experimental stage, will have the ability to remove humans from the battlefield altogether, both in the operational and decision-making sense. As he describes the trajectory of application and use for robotics, Ravi Mohan projects that robots will be engaged in a wide range of non-lethal activities including mine clearing and detection. Indeed, there has been a fairly robust use in this regard for at least a decade. Security applications come next such as patrolling base perimeters and the use of unmanned aerial robots (drones) to enforce no-fly zones and contested borders. In the final stage, robotic weapon will be given the technology and license to open fire on targets identified by their processors without permission or direction from a human in the loop.[19]

Robotics have responded to multiple-use scenarios, especially since 9/11 and the wars in Iraq and Afghanistan. While the most pronounced uses have been in Unmanned Aerial Vehicles (UAVs) including the Air Force's Predator and Global Hawk, the Army's Hunter, Shadow, Raven, and Wasp, and the Marine's Pioneer and Dragon Eye, robotics have also been designed to perform multiple uses in the Naval forces including a range of unmanned surface vehicles (USVs) and unmanned underwater vehicles (UUVs such as REMUS, Remote Environmental Monitoring Unit) which perform reconnaissance, perimeter defense, mine sweeping, and waterway clearance operations. There are also multiple adaptions for medical vehicles (REV, Robotic Evacuation Vehicle and REX, Robotic Extraction Vehicle), shipboard maintenance as well as logistics vehicles (the Mule) and mini-tanks (the ARV).[20]

Often referred to as second generation robotics, development of exoskeleton suits proceeds a pace as well. Military exoskeleton suits are designed to fit around the body of a dismounted soldier providing him or her with enhanced physical and mental capabilities including improved physical strength and endurance as well as the ability to see, hear and smell at a hyper level. These advantages will provide soldiers the ability to perform both combat and logistical tasks well beyond what the average soldier is capable of today.[21] Designed for multiple uses including prosthetic aids and as substitutes for heavy-lift machinery, exoskeletons will permit soldiers to carry more, run farther, and endure otherwise degrading weather and terrain conditions with minimal exertion. Examples of projects in the works include the Human Universal Load Carrier (HULC) and the Raytheon XOS exoskeleton, both of which seek to enhance the ability of soldiers to perform heavy lifting tasks while reducing orthopedic injuries.[22]

Part four: non-lethal weapons

Non-lethal weapons technologies span a wide array of disciplines and attempt to address the modern dilemma of security forces (police, soldiers, and soldiers acting as constabulary) who in dealing with entrenched and unruly adversaries of various kinds are often left with the Hobson's choice of projecting too little or too much force.[23] They attempt to address modern concerns regarding the interaction of state security forces and the public which are reflected in increased scrutiny of their activities, public diplomacy issues and ever-increasing humanitarian law concerns regarding proportionality and use of force generally. A fairly comprehensive definition speaks to their intentions rather than specifying specific technologies.

> 3.1 Non-lethal Weapons. Weapons that are explicitly designed and primarily employed so as to incapacitate personnel or materiel while minimizing fatalities, permanent injury to personnel, and undesired damage to property and the environment.
> > 3.1.1 Unlike conventional lethal weapons that destroy their targets principally through blast, penetration and fragmentation, non-lethal weapons employ means other than gross physical destruction to prevent the target from function.
> > 3.1.2 Non-lethal weapons are intended to have one, or both of the following characteristics:
> > > 3.1.2.1 They have relatively reversible effects on personnel or materiel.
> > > 3.1.2.2 They affect objects differently within their area of influence.[24]

Again, the technologies themselves do not define the category but rather their capabilities (or constraints) and their intention of use. Examples include sticky and slippery foam, various types of electric guns, often referred to in part as Tasers, Pepper Spray, Acoustic Rays, Directed Energy Heat Rays, Chemical Calmatives or Malodorants, Projectile Netting, Anti-materiel Biological and Chemical Agents and other miscellaneous non-lethal weapons such as electromagnetic pulse devices for disabling electrical systems, flash-bang and stinger grenades and low-kinetic-energy bullets.[25] There has been a good deal of uneasiness regarding defining these technologies as somehow separate and distinct from other forms of weapons systems inasmuch as all technologies are capable of lethality if improperly used.[26]

Part five: Cyber technology

Cyber technology, often referred to as information technology (IT), has created the most immediate change on the battlefield to date. Applications of emerging micro-chip-based technologies, especially advanced computers and communications systems, make it easier to find targets with precision, and kill the enemy with *smart* technology.

Precision Guided Missiles (PGMs), for example, were highly touted as a means to increase lethality while at the same time decreasing collateral damage, specifically civilian casualties. George and Meredith Friedman in their 1996 work, *The Future of War,* made the claim that the accuracy of PGMs may well change the moral calculus upon which munitions and targeting decisions are traditionally made in warfare. Even if war is ever a part of the human condition, the Friedmans argue, weapons such as the PGM, given their accuracy and the enhanced discretion with which they are used, may well end the slaughters of the 20th century and redefine the concept of collateral damage.[27]

Yet, there is a good deal more to cyber technology than merely the creation of efficiencies in weapons systems. Indeed, attempts to define cyberspace, the domain in which cyber technologies operate, has a history which can be characterized as contentious and changing. First of all, it is a physical place like the other domains of land, sea, aerospace, and outer space. It is set apart from the other domains by the fact that it is entered and used by the energies and properties of the electromagnetic spectrum of technologies. In cyberspace, electronic technologies are used to create, store, modify, exchange, compare, and exploit information. This is done through a networking of interdependent and interconnected networks using information-communication technologies.[28]

It is important to recognize the multiple dimensions and capabilities that this set of technologies incorporates. Daniel T. Kuehl describes the wide-ranging effects of cyberspace technologies, noting that networks and systems work simultaneously both in physical and virtual space within and outside geographic borders. Their users are clearly nation-states but also include individuals and transnational groups with seemingly no allegiance to governance regimes or best practices. There are, essentially, three separate yet interrelated dimensions which taken together define the global information environment. First, there are the physical platforms, systems and infrastructures which connect users with each other and the information they need to accomplish their multiple tasks. Second, there is the information itself, that vast amount of content which is stored, analyzed, and communicated digitally and electronically around the world real time. Content connectivity forms the basis for the enhancement of multiple activities. Users are simply able to accomplish more, more efficiently and in shorter time frames because of their ability to tap into this content. Finally, there is the dimension of human cognition itself, the ability or inability of human users to operate with the benefit of massive amounts of information. Inability to *keep up* with the quality and quantity of information can radically affect human behavior generally and decision-making specifically.[29]

Some of the characteristics of the cyber world bear mentioning. First, it is poorly regulated and extremely insecure from a national security point of view.[30] Second, it is *democratized* in the sense that the barriers for entry are low as is the ability to create applications for launching various types of cyber interventions (including attacks). Third, cyber technologies are not military technologies but rather are universal in their applications and the modern world's dependence on them.[31] These are virtues in the sense that they permit all manner of human endeavor including warfare to occur at a faster rate and with increased efficiency. They also comprise a set of vulnerabilities which argue for some sort of governance. The National Strategy to Secure Cyberspace summarizes these vulnerabilities: "[B]y exploiting vulnerabilities in our cyber systems, an organized attack may endanger the security of our Nation's critical infrastructures. The vulnerabilities that most threaten cyberspace occur in the information spaces of critical infrastructure enterprises themselves and their external supporting structures, such as the mechanisms of the Internet.

Lesser-secured sites on the interconnected network of networks also present potentially significant exposures to cyber attacks. Vulnerabilities result from weaknesses in technology and because of improper implementation and oversight of technological products."[32]

Notes

1. One study breaks the discussion down into "domains." Foundational sciences include information technology, synthetic biology, and neuroscience while application domains include robotics and autonomous systems, prosthetics and human enhancement, cyber weapons, and non-lethal weapons. Jean-Lou Chameau, William F. Ballhaus, Herbert S. Lin eds. *Emerging and Readily Available Technologies and National Security: A Framework for Addressing Ethical, Legal, and Societal Issues* (Washington D.C.: The National Academies Press, 2014), 31–32.
2. Daniel Ratner, Mark A. Ratner, *Nanotechnology and Homeland Security, New Weapons for New Wars* (Upper Saddle River, NJ: Pearson Education Inc., 2004), 8.
3. Ratner and Ratner, *Nanotechnology and Homeland Security,* 28.
4. Ibid., 7.
5. John Petersen and Daniel Egan, "Small Security: Nanotechnology and Future Defense," *Defense Horizons* 8 (March 2008) as cited in Armin Krishnan, *Killer Robots, Legality and Ethicality of Autonomous Weapons* (Burlington, VT: Ashgate Publishing Company, 2009), 85.
6. Patrick Lin and Fritz Allhoff, in Allhoff et. al. eds. *Nanoethics*, 11.
7. House Armed Services Committee. "Statement by Dr. Tony Tether, Director Defense Advanced Research Projects Agency, before the Subcommittee on Terrorism, Unconventional Threats, and Capabilities House Armed Services Committee, United States House of Representatives" (March 27, 2003) as cited in Moreno, *Mind Wars*, 11.
8. JASON, the MITRE Corporation, Human Performance, JSR 07 625, Study performed on behalf of the Office of Defense Research and Engineering Project no. 13079022 (March 2008), 12–13.
9. Moreno, *Mind Wars*, 116–132.
10. The President's Council on Bioethics, "Beyond Therapy: Biotechnology and the Pursuit of Happiness" (October, 2003), retrieved at http://bioethicsprint.bioethics.gov/reports/beyondtherapy.chapter-1..html, November 10, 2009.
11. Francis Fukuyama, *Our Posthuman Future, Consequences of the Biotechnology Revolution* (New York: Farrar, Straus and Giroux, 2002), 16.
12. Singer, *Wired For War*, 151.

13 Ibid.
14 Department of Defense, FY2009–2034 *Unmanned Systems Integrated Roadmap*, xiii.
15 Ronald O'Roarke, "Unmanned Vehicles for U.S. Naval Forces: Background and Issues for Congress," CRS Report for Congress, updated April 12, 2007, 1.
16 Department of Defense, FY2009–2034 *Unmanned Systems Integrated Roadmap*, xiii.
17 Armin Krishnan, *Killer Robots*, 2.
18 Marvin Minsky, *Semantic Information Processing* (Cambridge, MA: MIT Press, 1968), V.
19 Ravi Mohan, "Robotics and the Future of Warfare," Ravi Mohan's Blog [online] (December 13, 2007), retrieved at http://ravimohan.blogspot.com/2007/12/robotics-and-future-of-warfare.html, last retrieved October 10, 2010, November 25, 2009.
20 Max Boot, "The Paradox of Military Technology," *The New Atlantis, A Journal of Technology & Society* (Fall, 2006), 23–24. See also P.W. Singer, "Military Robots and the Laws of War," *The New Atlantis, A Journal of Technology & Society* (Winter, 2009), 31–34.
21 Liam Stoker, "Military Exoskeletons Uncovered: Ironman Suits a Concrete Possibility," *Army Technology Market & Customer Insight* (January 30, 2012), retrieved at http://ww.army-technology.com/featuremilitary-exoskeletons-uncoered-ironman-suits-a-concrete-possibility, April 29, 2014.
22 Mathew Ponsford, "Robot exoskeletons suits that could make us superhuman," CNN (May 22, 2013), retrieved at http://cpf.cleanprint.net/cf/cpf?action=print&type=filePrint=cnn&url=http%3A%2F%2Fwww.cnn.com, May 4, 2014.
23 David A. Koplow, *Non-Lethal Weapons, The Law and Policy of Revolutionary Technologies For the Military and Law Enforcement* (Cambridge: Cambridge University Press, 2006), 1–2.
24 Department of Defense Directive No. 3000.3, *Policy for Non-Lethal Weapons* (July 9, 1996).
25 Koplow, *Non-Lethal Weapons*, 14–28.
26 See generally, Neil Davison *"Non-Lethal" Weapons* (New York: Palgrave Macmillan, 2009), 1–10. "Paradoxically, despite increased research and development during the past 15 years, few 'non-lethal' weapons incorporating new technologies have actually been deployed on a large scale," 9.
27 George and Meredith Friedman, *the Future of War, Power, Technology, And American World Domination in the 21st Century* (New York: St. Martin's Press, 1996).
28 Daniel T. Kuehl, "From Cyberspace to Cyberpower: Defining the Problem" in Franklin D. Kramer, Stuart H. Starr, Larry K. Wentz, eds. *Cyberpower and National Security*. Center for Technology and National Security, National

Defense University (Washington, DC: Potomac Books, 2009), 28. The White House has used another definition:

> Cyberspace means the interdependent network of information technology infrastructures, and includes the Internet, telecommunications networks, computer systems, and embedded processors and controllers in critical industries.
>
> National Security Presidential Directive (NSPD) 54/Homeland Security Presidential Directive 23, "Cybersecurity Policy" (January 8, 2008).

29 Daniel T. Kuehl, "From Cyberspace to Cyberpower" in Kramer, et al. eds. *Cyberpower*, 28.

30 One recent commentator, for example notes:

> This principle [what precisely constitutes an armed attack or use of force in cyberspace] is described in multiple authoritative legal commentaries. But these can imply misleadingly that this consensus is global and unchallenged. In fact, China, Russia, and a number of like-minded nations have an entirely different concept of the applicability of international law to cyberspace as a whole, including the nature of conflict within it.
>
> Keir Giles with Andrew Monaghan, *Legality in Cyberspace: An Adversary View* (Carlisle, PA: Strategic Studies Institute, U.S. Army War College, March 2014), ix.

31 Wesley Clark, Peter Levin, "Securing the Information Highway, How to Enhance the United States' Electronic Defenses," *Foreign Affairs* (September/December, 2009), vol. 88, no.6, 4.

32 The White House, *The National Strategy to Secure Cyberspace* (Washington, DC: The White House, February 2003), xi.

3
Innovators and Consumers: The Culture of Innovation and Use of Military Technology in the 21st Century

Abstract: *A particular technological application makes its way to the battlefield for use in multiple ways. Innovation does not happen in a vacuum. There are important cultural, economic, social, and political influences which impact the push-pull process and determine what piece of technology is dedicated to a particular use and how it is used. The history of technological innovation must be understood in order to recognize the strengths and pitfalls of the process, especially in an era when mistakes can lead to unprecedented consequences.*

> Part one: Learning from experience: the intersection of technology and warfare
> Part two: The contemporary culture of military innovation

O'Meara, Richard Michael. *Governing Military Technologies in the 21st Century.* New York: Palgrave Macmillan, 2014. DOI: 10.1057/9781137449177.0007.

The research and development (R&D) of technology for military use, *co-option, innovation,* and *application,* is big; that is, it influences considerably the budget process, the relationship between the private and public sector, the relationship between institutions within the public sector, the economy generally, and relations between actors on the international stage. Exact numbers regarding government spending are difficult to get at due to the increasingly sophisticated and confusing manner in which they are reported and the fact that a fairly substantial portion of the investment by governments is classified. Further, technology innovation is often *dual-use,* that is, being performed by a wide range of civilian and military institutions, including research universities, commercial laboratories, government facilities, and individuals working in private facilities. Finally, since World War II, innovation culture has changed radically producing, what Philip Scranton has called, *technological uncertainty.*[1]

The intersection of technological innovation and warfare has a long history with multiple consequences.

Part one: Learning from experience: the intersection of technology and warfare

Because technology begets more technology, the importance of an invention's diffusion potentially exceeds the importance of the original invention.[2]

Jared Diamond notes that invention is less about creating new technologies out of whole cloth than it is about adapting ideas already in existence.

Often innovation arises from improving and deepening current technologies, using existing tools to find cheaper and more efficient ways to do old things. Sometimes innovation arises from borrowing ideas from different domains and applying them in new ways. Occasionally, a radical new innovation like electricity or the transistor comes along, making a whole generation of previously unthinkable technologies possible.[3]

Further, not all innovation creates radical changes in the way warfare is conducted. Major innovations challenge the way operations are conceived, the traditional relationships between combat arms and the identification of missions. Stephen P. Rosen concludes that such innovations have involved "…a new way of war, with new ideas of how the components of the organization relate to each other and to the enemy and new operational procedures conforming to those ideas. They involve changes in critical task[s], the tasks around which warplans revolve."[4]

It can be argued, with a fair degree of certainty, that technological innovation and its use on the battlefield comprise one of three or four important considerations which have influenced the ability of cultures generally to thrive or decline competitively in the course of human history. Indeed, the literature is robust in support of this proposition.[5] Other considerations such as demographics,[6] geography,[7] and ecology[8] must also take pride of place in the discussion regarding the ability of particular cultures to prosper. Diamond's question in this regard—*why did wealth and power become distributed as they now are, rather than in some other way?*—is relevant. Stated another way, for instance, *why weren't Native Americans, Africans, and Aboriginal Australians the ones who decimated, subjugated, or exterminated Europeans and Asians?*[9] These questions are seminal and to date the answers are not completely understood.

Clearly, though, technological innovation has had a huge impact. The ability of humankind to *co-opt, innovate, apply,* and *manage* (or regulate) technologies has made a difference. The history of humankind's reactions to these myriad advancements may contain some lessons regarding contemporary responses to emerging technologies and, therefore, bears reviewing.

As McNeill and others school us, technological diffusion can create change on a radical scale. Change occurs as new ideas, products or ways of doing things are introduced by people outside the community. In one form or another, these *new things* are attractive to those persons in the community with the power to accept challenges and make changes.[10] Diffusion has occurred in any number of ways and within a variety of contexts—commercial interaction, social interaction, conquest, migrations amongst others—and has not been consistent throughout history. Further, the rate of diffusion, for example, its speed and the willingness of groups to accept particular ideas and technologies has varied widely as well. There appear to be a number of reasons for this phenomenon. Diamond, for example, concludes that human ingenuity is, perhaps, the least important factor. He hypothesizes that geography, perhaps more than any other condition, has affected both the speed of diffusion and political will most.[11] His conclusions regarding the inability of the Incan civilization and civilizations in Sub-Saharan Africa to remain competitive with Europeans, for example, are instructive. His answer revolves around the technologies of guns, steel weapons, and the use of horses as well as maritime technologies, the centralized organization or European states and communication not to mention the introduction of infectious diseases. Diamond goes to great lengths to demonstrate that Europe's colonization of Africa, for example, had nothing to do with differences in African peoples themselves, as generations of white historians and social scientists have assumed, but rather is due to accidents of geography especially the continent's different areas, axes, and suites of wild plants and animals.[12]

There are other considerations as well which may be deemed *cultural*, although their relationship to geography and ecological circumstance bears remembering. Europe, for example, was blessed early-on with the use of the wheel and the domestication of animals.[13] These technological innovations, amongst many others, migrated across Eurasia fairly easily, and contributed considerably to the ability of Europeans to communicate, transport goods and ideas, and otherwise benefit from a relatively free (unregulated) exchange of innovations. On the other hand, the continent remained until very recently a fragmented space, chock full of competing political, social, and economic centers of power, and eager to a fault to obtain an advantage, one against the other.[14] The separation of the Roman Church from the other political arrangements during the Middle Ages, unlike in Byzantium and China, for example, insured that no overarching regulation of ideas and technologies could occur.[15] It is generally accepted that this set of political and social arrangements created a culture of competitiveness only aggravated by the practice of Enlightenment individualism and capitalism and the peculiarism of nationalism. Indeed, even the worst effects of these arrangements—multiple and massive warfare from the 17th through the 20th centuries—contributed to an environment of technological innovation seen nowhere else on the planet.[16]

The West, then, has exhibited an unparalleled ability to *co-opt* technology and bend it to its use, *adaption*, through its history. As Alex Roland concludes, while technology does not determine their use and the ultimate intended and unintended consequences thereof, certain cultures have exhibited a willingness to walk through the *open door* of technological possibility more than others. The West has been one of those cultures.[17]

There are multiple examples of technological *co-option* in human history.[18] Indeed, there is a good deal of evidence that transformative technologies arise, not from "thinking outside the box," ingenuity and genius and so on, but, rather, from *combinational evolution*, a process whereby technologies are put together from other technologies, often to solve the unintended consequences of former technologies.[19] Two are particularly instructive and make the point: the adoption of the stirrup in 7th-century Europe and gunpowder in the 15th century.

Those who dominated military culture in the Early Middle Ages in Europe were the inheritors of at least two military traditions with which to confront the extremely serious security dilemma of bands of mounted cavalry swarming from the steppes of the East and across the Pyrenees. The Roman tradition, characterized primarily by well-organized and disciplined units of infantry, was difficult to continue in a world with little ability to marshal, train, and maintain large groups of professional warriors.[20] Dismounted infantry also worked poorly against mounted cavalry, capable of rapid movement and the ability to retreat at will. Nor did its emphasis on unit cohesion speak to the Germanic tradition of individual warfare and its reverence for individual combat, reward, and reputation.[21]

The idea of the stirrup diffused into Europe from multiple sources over a fairly long period of time, not the least of which were the Byzantine Empire and the Saracen advances through Spain in the 8th century A.D. Some theorize that the stirrup was introduced by the Lombards and Avars;[22] others credit the Franks with its *co-option* and general use. The point here is that this technology, cheap and readily adaptable to a relatively disorganized military organization of minimal numbers seeking mobility against mounted cavalry, affected the balance of power considerably. Had the Franks and others continued to ride to battle without the stirrup and the heavy armor and lance which the stirrup permitted, they would have continued to dismount to fight. Further the geographical reach of Charlemagne and others would have been severely limited as well. And their ability to close with and destroy mounted cavalry would have continued to be negligible. Some have argued that this piece of technology (and more importantly the ability to *adapt* it for uses on European battlefields) changed the entire socio-economic, political, and cultural history of Europe. Lyn White, in a famous 1966 article, speculated that the stirrup represents one of the simplest yet most catalytic inventions in human history, creating not only new ways of projecting force but new ways of organizing and ordering society. This new mode of warfare made it possible for a new class of warrior to emerge, one mounted, carrying heavy armor and capable of ranging well outside traditional borders. Economic and social arrangements developed to support him as he saw to the security of Western Europe. "The Man on Horseback, as we have known him during the past millennium, was made possible by the stirrup."[23]

Others disagree with Lynn's conclusions.[24] No one, however, denies the fact that *co-option* and *adaption* of this technology changed the paradigm of war-making in Europe, in multiple ways for centuries.

The history of gunpowder technology is, perhaps, a more obvious example of the effects of *co-option* and *adaption*. Gunpowder was not new to human history in the Middle Ages. Greek fire first appears as part of Byzantine technology in the 7th century. It was discharged in liquid form in order to serve as an incendiary agent against wooden structures in siege and naval warfare. In a sense it was not gunpowder at all. Its history, however, demonstrates the process of adaption. Keegan points out the quixotic manner in which discoveries find their way into regular and important applications. The Chinese early on, perhaps as early as 950 A.D., discovered that various mixtures of sulfur, saltpeter, and charcoal produced an incendiary compound which was used for semi-magical rituals in Taoist temples. There is considerable disagreement regarding Chinese use of these substances in warfare before the 13th century, but there is no evidence that they fashioned cannon of any kind. It appears, therefore, to represent an insignificant piece of technology until it travelled into the west.[25]

Its *co-option* and then *adaption* to European and then global projects appear to have made all the difference. First, Europeans applied its use to a whole range of stand-off weapons which challenged the defensive castle warfare of the fifteenth and sixteenth centuries. Again, standoff weapons, those which permitted the projection of violence without the necessity of face-to-face contact, were not unknown before the age of gunpowder. The catapult and other such contrivances have a history which long predates even the Greeks and is global in its applications. The long-bow in Europe was particularly influential in operations between the English and French as well.[26] Cannon, created in conjunction with bell-foundry technology already well developed in Europe, were mobile, reasonably accurate, and immediately decisive in a wide range of operations. Christopher Duffy, for example, describes the process of adapting gunpowder, bell-foundry technology, and the precision of muzzles as they impacted the siege of Constantinople in 1453. Initially, cannon or *bombards* fired a stone ball from a wooden platform that was moved by cart as needed.

This proved inefficient and was replaced by a slender, bronze cast tube of approximately eight feet, whose proportions were calculated to absorb the shock from the muzzle. This new configuration fired wrought iron balls, heavier than their stone equivalents and therefore considerably more destructive.[27]

Their expanse hastened the shift from independent war-makers to more centralized polities and forms the basis for at least part of the reason for the formation of the modern state.[28] The progression of gunpowder technology into the area of individual use—the musket and then the rifle—are of equal and, perhaps, more importance. The wide-scale arming of individual infantrymen with muskets and ring bayonets required considerable improvement in methodologies of finance, regimens of discipline, and interactions between commerce and the state. Technological innovation spread across the entire range of organizational activities with far reaching results.

European kings and captains had clearly accepted the idea that improvements were always possible. An efficient information network utilizing printed texts as well as word-of-mouth espionage, and commercial intelligence, spread data about enemy intentions and capabilities, new technologies, and new tactics across the length and breadth of western Europe. As a result, by the end of the Thirty Years War, European armies were no longer a mere collection of individually well-trained and bellicose persons, as early medieval armies had been, nor a mass of men acting in unison with plenty of brute ferocity but no effective control once battle had been joined, as had been true of the Swiss pike men of the 15th century. Instead, a consciously cultivated and painstakingly perfected art of war allowed a commanding general, at least in principle, to control the actions of as many as 30,000 men in battle. Troops equipped in different ways and trained for different forms of combat were able to maneuver in the face of an enemy. By responding to the general's command, they could take advantage of some unforeseen circumstance to turn a stubbornly contested field into a lopsided victory. European armies, in other words, evolved very rapidly to the level of the higher animals by developing the equivalent of a central nervous system, capable of activating technologically differentiated claws and teeth.[29]

Adaptions of these technologies have created the categories of weapons systems which continue to dominate the contemporary battlefield.

From a global perspective, the *adaption* of gunpowder technologies to naval science in the 15th century led to the ability of the west to dominate naval warfare—and therefore the global commons—and must be considered, therefore, of equal if not larger significance. Early on, the ability of Atlantic fleets to create powerful platforms for cannon almost immediately changed the balance of power throughout the oceans of the world and wiped out millennia of traditional naval tactics. These technologies and their accompanying tactics, according to McNeil, changed the balance of power in the Mediterranean Sea as well as in the Indian Ocean almost immediately. In less than a generation, the Portuguese realized considerable success against the Muslim navies of the Indian Ocean. In 1509, for example, off the port of Diu, the Portuguese were able to stand off from numerically superior groups of ships (200 yards), avoid the tactic of close combat, boarding and melee, and destroy the enemy's platforms with considerable success.[30]

Of considerable interest is the failure of other cultures to *adapt* the same technologies in order to compete with the voracious proclivities of the West. China, after all, is often credited with inventing gunpowder, the stirrup, and the crossbow, and the gates of Constantinople—and hence the last vestiges of Roman Empire—were breached by the Ottoman Turks in 1453 with the use of heavy cannon.[31] The answer appears to lie in the spheres of culture and politics. Here, it can be argued, are the first seeds of *regulation* of the *genie*. The irony, of course, is that this *regulation* had the disastrous consequence of rendering these cultures, Chinese specifically and Islam generally, incapable of competing with what is often referred to as *modernity*.

Modernity, of course, is a much argued concept which is often defined in Euro-centric terms, especially when it speaks to the importance of the scientific and commercial revolutions, the Enlightenment, the Industrial Revolution, and globalization. Clearly though, it is bound up at least in part with issues of *co-option* and *adaption* of technology. Richard Hooker provides one definition: "[M]odernity is simply the sense or the idea that the present is discontinuous with the past, that through a process of social and cultural change (either through improvement, that is progress, or through decline), life in the present is fundamentally different from life in the past. This sense or idea as a world view contrasts with what I will call tradition, which is simply the sense that the present is continuous with the past, that the present in some way repeats the forms, behavior, and events of the past."[32] The concept of modernity speaks not only to the record of technological innovation but also to the idea that new things are not only possible but their discovery is appropriate and adds value to the quality of life of societies. There is an assumption here that change is both inevitable and worthwhile. The concept assumes the inevitability of change as well as its utility. There are conations of progress, rationality, and purposeful action, universal norms and promises of a better life which are inferred as well.[33]

China, for example, has been characterized for centuries as culturally conservative, defined by a Confucian belief in order and the centrality of government, originating in part as a result of the social and economic implications of rice paddy culture and a near constant threat of anarchy. The commercial and military professions have traditionally been seen as lesser occupations and did not hold prestige as did bureaucratic service and landholding.[34] Further, what can be described as the Chinese *way of war*, as enunciated by Sun Tzu and others, emphasized avoiding battle except with the assurance of victory, disfavoring risk, seeking to overawe an enemy by psychological means, and using time rather than force to wear an invader down. There is a good deal of evidence that this cultural proclivity continues at least in part today. The US Department of Defense, for example reports:

> ...Chinese strategists and analysts occasionally cite guidance from former paramount leader Deng Xiaoping in the early 1990s: "observe calmly; secure our position; cope with affairs calmly; hide our capacities and bide our time; be good at maintaining a low profile; and never claim leadership". This guidance reflected Deng's belief that China's foreign policy and security strategy had to reinforce its core national interest of promoting domestic development by avoiding foreign risk, high-profile international engagement and provocations, or pretenses of international leadership.[35]

These proscriptions are all profoundly anti-western in philosophy and practice and hardly encourage a freewheeling competition of ideas and innovation. They have informed the Chinese way of war for centuries. As Keegan reports, the Chinese *way of war* emphasized the ideal of rationality, continuity, and maintenance of institutions which subordinated warrior practice to the constraints of law and custom. The most persistent character of the way Chinese culture approached the project of war was the emphasis on moderation designed to preserve cultural forms. There is little room here for foreign conquest or internal revolution, and therefore minimal need for the cut throat competition of innovators seeking to create and sell new military weapons and equipment to commanders always seeking an advantage.[36]

Again, according to Keegan, these restraints constituted a very particular method of arms control practiced not only in China but in Japan as well where elites in the 16th century chose to forgo the use of the known and developed technology of musketry and cannon in favor of traditional weapons, cavalry, bows, swords and so on.[37] These cultural practices bore heavily on the ability of Asian culture to compete with European practices of innovation and adaption. Keegan concludes, "[T]he western world, by forsaking arms control, embarked on a different course, which resulted in a different form of warfare that Clausewitz said was war itself, a continuation of politics, which he saw as intellectual and ideological, by means of combat, which he took to be face-to-face, with the instruments of the Western technological revolution, which he took for granted."[38]

Finally, there is the issue of command economies and their relationship to innovation. Unlike China, Constantinople, and Islam generally, Europe never succeeded in melding church and state into one institution. Technological innovation proceeded in these cultures [Constantinople and China], when it did, as a result of the decisions of one source of power whose agendas were mostly concerned with maintenance of order and the status quo.[39] The ability to *regulate* the methodologies of warfare, then, was considered one of the virtues of governance.

Islam, too, has had a tradition of restraint in war-making that has proven beneficial, at least in part, over the ages. Despite its reputation for conquest and its early successes in this regard, the theology of Islam schools is a prohibition against war and pitting one Muslim against the other. Keegan argues that this prohibition led to the formation of a specialist and subordinate class, which freed the majority from the vicissitudes of military obligation and permitted the pious to devote their lives to "...the 'greater' rather than the 'lesser' aspect of the injunction to wage holy war, 'the war against self.'"[40] Command politics and economies and emphasis on a single theological way of doing things has stifled the freedom to innovate and created a dependence on Western technologies and practices.[41]

The above short history demonstrates, in part, that technological innovation is nothing new to humankind, indeed it may well form part of the definition of what it means to be human. Certain conclusions may be drawn from past experience:

1   There are many things which affect the ability of a culture to *co-opt* and *adapt* new ideas, especially technological ideas. Some of these

are the existence of social, political, and economic environments where new technology is prized and rewarded;

2 Innovation is often, if not always, the product of *co-option* and *adaption*. While all innovation is not the result of providing a response to an immediate or emergent challenge, immediate and emergent challenges often spur perceived *fixes*, all of which have unintended and often far-reaching consequences; and

3 There has been a good deal of *regulation* regarding technological innovation which has attempted to control the means available to project violence, restricting its use to certain groups and certain practices within cultures. While this regulation has been beneficial, it has also had the effect of leaving cultures open to conquest by other more unregulated cultures.

Part two: the contemporary culture of military innovation

Philip Scranton explains that a number of perceived emergent requirements since World War II have resulted in two shifts in the way innovation culture has organized. These requirements included organization around the perceived challenges of the Cold War which established multiple bipolar, global, and political-technical competitions. The United States established a large, permanent-standing army fed by the draft and now a mature reserve force, a parallel intelligence arm of the executive branch and an emphasis on new and rapidly changing technological innovations for warfare. This new environment of innovation has created multiple unintended consequences in the manner in which an idea moves through the process of adaption and use.

> The first shift during the Cold War affected a number of industrial fields where urgent demand, funded by rival military establishments, propelled what I'm calling experimental development of highly complex, yet workable devices—despite insufficient usable or relevant science... Second, a pattern of continuous innovation along many of these trajectories [metallurgy, fluid dynamics, combustion, etc.] entailed that design changes multiplied and user expectations altered at rapid rates. This meant that no technologically stable platform could be realized so that iterations of use could squeeze out faults and allow remediation. In essence, continuous redesign in the context of incomplete (or underdeveloped) science created durable or, in Karl Weick's terms,

"permanent" technological uncertainties. Neither military nor commercial rivalry permitted a freezing of designs that in turn could allow learning from failures to generate deep knowledgeability and condition a technological stabilization, as seems to have happened so often in earlier generations. In consequence, stochastic failures followed redesigns in irregular order.[42]

These shifts have informed the environment in which innovation occurs today. First, both cost expectations and planned schedules for development and delivery regularly prove unreliable. Constant complaints regarding *fraud, waste, and abuse* in the procurement system, while often justified, routinely occur as a result of a built-in dilemma regarding the manner in which innovation is produced—"...getting a novel device on time and on budget could easily mean getting a devise that lacked innovation, was obsolescent at first use, worked unreliably, or all three."[43] Second, constant redesign plays havoc with the management of logistics, maintenance, and supply. Third, redesign and operational technological uncertainty constrains operational deployment of military technologies because user's training and experience is routinely degraded by the introduction of new redesigns and fixes to apparent problems, and entire units, configured on the basis of the use of particular technologies are rendered non-deployable as a result of redesigns and fixes. Fourth, a good degree of what Scranton refers to as *political uncertainty* results in that rumor of new developments has had a way of affecting the internal politics of the services as they fight for mission dollars and relevance. Further, political wrangling regarding the entire process, its constant morphing from support for one program and then the next and the various interest groups poised always to take advantage of the *next big thing*, contributes to an inability to see confidently into the future and plan accordingly.[44] Finally, perhaps most important for the purposes of this discussion, the rapidity of innovation curtails analysis of the intended and unintended consequences which may occur. The *genie* is let loose on the world without study or reflection, certain only in the inevitability of his coming.

This competition is of a particular stripe as well. Before the end of the Cold War, it has been argued, the nature of arms races between the super powers—Britain, France, Japan, Germany, the US, and the USSR—were primarily *quantitative*, that is, everybody pretty much had the same weapons (dreadnoughts, tanks, aircraft carriers, bombers, nuclear warheads); the question was how many were operational in the various arsenals. Since the end of the Cold War, however, the competition is more about new technology, *qualitative* superiority, and the competition includes a race to outspend others.⁴⁵ Further, innovation speaks not only to creating offensive weapons-first use-but also combating technologies created by others. Thus, one argues, even if a state forgoes the use of a particular weapon, a particular virus for example, it must continue its research in order to defend against another state that may—or may not—be so inclined. DARPA, for example, emphasizes research in nine strategic areas, one of which is referred to as *bio-revolution*. DARPA reports that there is a requirement for *keeping-up* especially in the areas of vaccine and pathogen development.

> Developing defences against biological attack poses daunting problems. Strategies using today's technologies are seriously limited. First, it is nearly impossible to predict what threats might emerge in two decades, particularly engineered threats. Second, from the moment a new pathogen is first identified—either a weapons agent or a naturally emerging pathogen—today's technology requires at least 15 years to discover, develop, and manufacture large quantities of an effective therapy...
>
> DARPA's programs have begun to transition technologies to US Government agencies and commercial industries that will enable vaccine discovery to potentially occur orders-of-magnitude faster than we can make happen today, and in population-significant quantities.⁴⁶

Finally, the intersection between bureaucracies, civilian institutions, and individuals, all competing for research and development (R&D) dollars, contributes to internal arms races in and of themselves. There is considerable competition within the various branches of the military and the laboratories which support them for research and development dollars. This competition and the bureaucratic infrastructure that has been created to foster it insure sustained internal competition in order to maximize weapons efficiencies, obtain dramatic results, and justifications for their continued existence.⁴⁷

There is the added temptation to political as well as economic corruption that is engendered as a result of this competition as well. President Dwight D. Eisenhower recognized this fact as early as 1961. His warning is worth repeating here.

> The conjunction of an immense military establishment and a large arms industry is now in the American experience. The total influence—economic, political, even spiritual—is felt in every city, every State House, every office of the Federal Government... We must guard against the acquisition of unwarranted influence, whether sought or unsought, by the military industrial complex. The potential for the disastrous rise of misplaced power exists and will persist.⁴⁸

Nor is all this innovation cheap. Some numbers, while hardly exact (another consequence of *technological uncertainty*?) are instructive. The United States Defense budget request for FY 2010 was in the area of $3.6 trillion. R&D spending is well over $78.6 billion. This does not include R&D spending which is classified, nor does it include R&D spending by other agencies which impact on military innovation.[49] China's R&D spending since 1995 has increased at an annual rate of 19 percent to reach $30 billion in 2005, the best figure that the Department of Defense could provide in its report to Congress in 2009. Given the continued economic growth of China in the interim, the lack of transparency regarding Chinese budgetary matters, and the labyrinth of intersecting private and public sector institutions, this figure must be assumed to be much higher.[50]

Finally, there is the issue of the government's role in innovation. While a good deal of the technological innovation which finds its way into military use comes from the civilian sector as either *dual-use* technology or technology which is adapted after the fact for military use, there is a fairly pervasive government footprint as well. Indeed, according to Barton C. Hacker, since World War II, US military culture has operated on the assumption that better technology means victory and, therefore military-technological innovation has become, to a significant degree, an end in itself.[51] There are multiple research projects of various sizes and charters which concern themselves with the short-term requirements based needs of commanders on the battlefield throughout DOD. Sitting above these is DARPA, an organization of approximately 150 individuals who are solely concerned with *radical innovation for national security*.[52] Born of the paranoia which surrounded the Soviet Union's Sputnik operation in space in 1957, DARPA's mission is "to prevent technological surprise for us and to create technological surprise for our adversaries."[53] It is the first to recognize that technologies which have found their way into military use were not developed initially as a result of military doctrinal requirements but rather came from adaption of civilian technological advances. Most of the weapons which have come to define modern combat in the 20th century, for example, the airplane, tank, radar, jet engine, helicopter, electronic computer, were created out of technological advances which were neither foreseen nor requested by the military as part of a doctrinal shift. DARPA has recognized that other, more contemporary technologies such as unmanned systems, stealth, and global positioning systems (GPS) as well as computer technologies, fall within the same category.[54]

DARPA looks well beyond the commander's requirement in most cases and emphasizes research... "the Services are unlikely to support because it is risky, does not fit their specific role or missions, or challenges existing systems or operational concepts."[55] DARPA receives approximately 25 percent of the DOD Science and Technology budget (almost $14 billion). Thus, DARPA's funding goes to long-range visionary innovation while the remainder of the DOD Science and Technology budget is devoted to product improvement and near-term requirements based projects.[56] The majority of DARPA investments (approximately 98 percent) go to organizations outside DARPA, primarily universities and industries. The purpose is to abet the outside institution's effort to create a technological innovation in which industry becomes sufficiently comfortable to invest its own money as it goes forward to propose it to a DOD user. DARPA employees are not career bureaucrats but move in and out of industry and academia during the course of a career. DARPA believes this fosters a culture of collegiality and innovation without the constraints of parochialism normally associated with the decentralized innovation organization of the agencies.[57] Its culture, then, is to free itself of all constraints in order to insure its ability to find *the next best thing*— radical innovation for national security.

All of this has occurred in a political environment which explains, some would argue mandates, the need for rapid innovation. Since World War II, the Korean War, and especially Vietnam, the American public has eschewed the concept of mandatory public service, has underwritten an expensive volunteer force for this purpose, and has demonstrated very little interest in participating in foreign policy projects led by the military. Demographically, the pool of volunteers is dwindling and there is a perceived belief—acted on by politicians generally—that the public has no tolerance for death and maiming on the battlefield—either side—no matter what the cause.[58] Technology has been and will continue to be proposed as the *fix* for these conditions. The political aspect of this innovation culture is described by Singer as fairly radical, de-emphasizing the organization of soldiers in support of weapons systems and units. Now, individual soldiers, armed with the mobility and firepower of traditionally formed large units, can operate alone or in small groups, separately empowered and unsupervised. Marine Lieutenant General James Amos optimistically posits that soldiers serving in tiny squads, commanded by a sergeant or lieutenant, could hold down hostile cities of 100,000 or more. New military technologies turn on their head not only the efficiencies of units but also the traditional way in which they organize, train, and are led.

[H]aving small units packing such punch would also change the way a nation mobilizes for war. Fewer soldiers would seem to be needed for the same task and a nation with technologically super-empowered soldiers might make it easier to strike quickly or covertly. If there were smaller numbers of troops in the field, it would also eliminate the need for a huge logistical support structure. Ultimately, described by one set of military analysts, "What we are seeing is the end of the G.I. The G.I., the stamped government issue [GI] interchangeable warrior, becomes obsolete when masses of men are no longer required to fight wars.[59]

This environment, then, is driven by multiple factors from the top and the bottom. It is generally decentralized but funded in one form or another by massive amounts of money from the central government, which is itself driven by the need to place technology of all kinds between itself and the people it serves. The system is vast, unorganized, and like many aspects of the globalized 21st century, ungoverned and as yet ungovernable.

Notes

1 Philip Scranton, "The Challenge of Technological Uncertainty," *Technology and Culture*, vol. 50, no. 2 (April 2009), retrieved at http://muse.jhu.edu.proxy.libraries.rutgers.edu/journals.technology_and_culture/v050/50, November 20, 2009.
2 Jared Diamond, *Guns, Germs, and Steel: The Fates of Human Societies* (New York: W.W. Norton & Co., 1999), 52.
3 Ibid., 258.
4 Stephen P. Rosen, "New Ways of War: Understanding Military Innovation," *International Security* (Summer, 1988), 134.
5 *See,* for example, Charles Singer ed. *A History of Technology* (Oxford: Clarendon Press, 1954–84, vols 1–5); Donald Cardwell, *The Fontana History of Technology* (London: Fontana Press, 1994); Arnold Pacey, *Technology in World Civilization* (Cambridge: MIT Press, 1990); Trevor Williams, *The History of Invention* (New York: Facts on File, 1987); R. Stephen Bull, *Encyclopedia of Military Technology and Innovation* (Portland, OR: Greenwood Pub, 2004); Victor Hanson, *The Western Way of War, Infantry Battle in Classical Greece*, 2nd edn (Berkley: University of California Press, 1989); Christopher Duffy, *The Military Experience in the Age of Reason* (New York: Barnes & Noble, 1997); Michael Howard, *War*

in Human History (Oxford: Oxford University Press, 1976); David Kahn, *Seizing the Enigma, The Race to Break the German U-Boat Codes, 1939–1943* (New York: Barnes & Noble, 2009); John Keegan, *The Price of Admiralty, The Evolution of Naval Warfare* (New York: Penguin Group (USA), 1990); McNeill, *The Pursuit of Power,* and *The Rise of the West (Chicago:* University of Chicago Press, 1992); G. Parker, *The Military Revolution, Military Innovation and the Rise of the West, 1500-1800,* 2nd ed (Cambridge: Cambridge University Press, 1996); Noel Perrin, *Giving Up the Gun, Japan's Reversion to the Sword, 1543–1879* (Boston: D.R. Godine, 1988); William Reid, *Arms through the Ages* (New York: Harper Collins, 1976); David Showalter, *Railroads and Rifles, Soldiers, Technology and the Unification of Germany* (Hamden, Conn.: Archon Books, 1975); Ian Brouma, *Inventing Japan* (New York: Modern Library, 2004); Martin L. Van Creveld, *Technology and War, From 2000 B.C. to the Present,* rev. edn (New York: The Free Press, 1991); Bernard Lewis, *What Went Wrong? Western Impact and Middle Eastern Response* (Oxford: Oxford University Press, 2002); Charles C. Mann, *1491: New Revelations of the Americas Before Columbus* (New York: Vintage Books, 2006); Diamond, G*uns, Germs and Steel;* and *Collapse, How Societies Choose to Fail or Succeed* (New York: Penguin Books, 2006).

6 See, for example, P.M.G Harris, *The History of Human Populations: Migrations, Urbanizations and Structural Change,* vol. 11 (Westport, CN: Praeger Pub, 2003); Massimo Livi-Bacci, *A Concise History of World Population,* 4th edn (Oxford: Blackwell Pub, 2007).

7 See generally Diamond, *Guns, Germs and Steel.*

8 Diamond, *Collapse.*

It has long been suspected that many of those mysterious abandonments the Anasazi and Cahokia within the boundaries of the modern US, the Maya cities in Central America, Moche and Tiwanaku societies in South America, Mycenaean Greece and Minoan Crete in Europe, Great Zimbabwe in Africa, Angkor Wat and the Harappan Indus Valley cities in Asia, and Easter Island in the Pacific Ocean] were at least partly triggered by ecological problems: people inadvertently destroyed the environmental resources on which their societies depended... The processes through which past societies have undermined themselves by damaging their environments fall into eight categories, whose relative importance differs from case to case; deforestation and habitat destruction, soil problems (erosion, salinization, and soil fertility losses), water management problems, overhunting, overfishing, effects of introduced species on native species, human population growth, and increased percapita impact of people. 6.

9 Diamond, *Guns, Germs and Steel.*

10 McNeill, *A History of the Human Community,* xiii. Anthropologists refer to this phenomenon as *cultural diffusion,* and it appears to apply to all manner of human endeavors such as religion, economics, political organization, the

exchange of human ideas generally and most importantly for the purpose of this discussion technological innovation.
11 Diamond, *Guns, Germs, and Steel*, 426.
12 Ibid., 182.
13 Ibid.
14 Yale H. Ferguson and Richard W. Mansbach, *Remapping Global Politics, History's Revenge and Future Shock* (Cambridge: Cambridge University Press, 2004), 230–32; Richard Langhorne, *The Coming of Globalization: Its Evolution and Contemporary Consequences* (New York: Palgrave Macmillan, 2001).
15 McNeil, *The Pursuit of Power*, 68–70.
16 Keegan, *A History of Warfare*, 390.
17 Alex Roland, "Presentation Notes at the Teaching the History of Innovation Workshop," published in *Footnotes, Foreign Policy Research Institute*, retrieved *at* http://www.fpri.org/footnotes/1402.200902.roland.wartechnology.html, November 11, 2009. "The open door is a powerful conceptual tool for thinking about all technology, especially military technology. It adds what most accounts of technological innovation lack: human agency," 4.
18 W. Brian Arthur, *The Nature of Technology, What It Is and How It Evolves* (New York: The Free Press, 2009).
19 McNeil, *A History of the Human Community* xiii.
20 David Grossman, "Evolution of Weaponry, A Brief Survey of Weapons Evolution. The Roman System." *Killology Research Group*, 1999, retrieved at http://www.killology.com/art_weap_sum_roman.htm, November 23, 2009.
21 John Sloan, "The Stirrup Controversy" posted on discussion list medieval@ukanvm.cc.ukans.edu on October 5, 1994, as part of the thread "The Stirrup Controversy," retrieved at http://www.fordham.edu/halsall/med/sloan,html, November 23, 2009.
22 David Edge, John M. Paddock, *Arms and Armor of the Medieval Knight* (New York: Crescent Books, 1988).
23 Lynn White, Jr. *Medieval Technology and Social Change* (Oxford: Oxford University Press, 1966).
24 For a discussion of the multiple disagreements with White's thesis see Sloan, "The Stirrup Controversy" posted on discussion list medieval@ukanvm.cc.ukans.edue on October 5, 1994, as part of the thread "The Stirrup Controversy," retrieved at http://www.fordham.edu/halsall/med/sloan,html, November 23, 2009.
25 Keegan, *The History of Warfare*, 319.
26 Captain Anton, "A Short History of the English Longbow." *Archers of Ravenwood*, retrieved at http://www.archers.org/default.asp?section=History&page=longbow, November 23, 2009; Keegan, *The Face of Battle* (New York: Penguin Group (USA), 1978).

27　Christopher Duffy. Siege Warfare: The Fortress in the Early Modern World, 1494–1660 (London: Routledge, 1996), 8–9.
28　Philippe Contamine, ed. War and Competition Between States (Oxford: Oxford University Press, 2000); Patrick Carroll, Science, Culture, and State Formation (Berkley: University. of California Press, 2006).
29　McNeil, The Pursuit of Power, 123–24.
30　Ibid., 99–100.
31　Roger Crowley, "The Guns of Constantinople," HistoryNet.com, retrieved at http://www.hstorynet.com/theguns-of-constantinople.htm, November 24, 2009; Steve Runciman, The Fall of Constantinople, 1453 (Cambridge: Cambridge University Press, 1965). Ironically the cannon technology that ultimately defeated Constantinople had been offered first to the Emperor by a Hungarian inventor. The Emperor found it useful but too expensive.
32　Richard Hooker, "Enlightenment Glossary, World Civilizations, 1996," retrieved at http://www.wsu.edu:8080/-dee/GLOSSARY/MODERN.HTM, November 24, 2009.
33　Thomas J. Misa, Philip Brey, Andrew Feenber, eds. The Compelling Tangle of Modernity and Technology (Cambridge, MA: MIT Press, 2004). 5.
34　McNeill, A History of the Human Community, 331.
35　Office of the Secretary of Defense, "Annual Report to Congress, Military Power of the People's Republic of China, 2009" (2009).
36　Keegan, The History of Warfare, 388.
37　Stephen Mirillo, "Guns and Government, A Comparative Study of Japan and Europe," Journal of World History. vol. 6, no. 1 (1995).
38　Keegan, the History of Warfare, 390–91.
39　The history of China, for example, is characterized by spurts of technological innovation and adaption, driven primarily by the government. In 1436, the Emperor issued a decree effectively shutting down the seagoing industry, closing shipyards and making it illegal to promote commerce overseas. This is especially significant since by then China had a growing naval commercial industry which routinely traded in the Indian Ocean and had reached the coast of Africa. McNeill, Pursuit of Power, 44–46.; Louise Levanthes, When China Ruled the Seas: The Treasure Fleet of the Dragon Throne, 1405–1433 (Oxford: Oxford University Press, 1994).
40　Keegan, A History of Warfare, 389. See also Sayyid Imam al-Sharif, Rationalizing Jihad in Egypt and the World, 2007 as cited in Jared Brachman, "Al Qaeda's Dissident," Foreign Policy, Special Edition, 2009, 40, in which he argues that Al Qaeda's use of mass violence against fellow Muslims is in violation of Islamic law. This is especially relevant, according to Brachman, because Sharif is one of the founders and original ideologues of the movement.
41　Bernard Lewis, What Went Wrong? 159.
42　Scranton, "The Challenge of Technological Uncertainty," 3.

43 Ibid., 4.
44 Ibid.
45 Alia Lamaadar, "War and Peace from Weapons Technology: Examining the Validity of Optimistic/Semi-Optimistic Technological Determinism." *The McGill Journal of Political Studies* (2003–04), 5.
46 Department of Defense, "Defense Advanced Research Projects Agency, Strategy Plan," May 2009, 37 and 39.
47 Lamaadar, "War or Peace from Weapons Technology," 5.
48 Farewell Address of President Dwight D. Eisenhower, January 1961.
49 Patrick J. Clements, "Research and Development in the FY 2010 Defense Budget," *Budget Insight, Stimson Center Blog*, November 3, 2009, retrieved at http://budgetinsight.wordpress.com/2009/11/03/research-and-development-in-the-fy-2010-defense-budget/, November 20, 2009.
50 Office of the Secretary of Defense, "Annual Report to Congress." 31–35.
51 Barton C. Hacker, *American Military Technology: The Life Story of a Technology* (Baltimore: Johns Hopkins University Press, 2006), 165.
52 Ibid., 1.
53 Ibid.
54 Ibid., 3, quoting John Chambers, ed., *The Oxford Companion to American Military History* (Oxford: Oxford University Press, 1999), 791.
55 Hacker, *American Military Technology*, 5.
56 Ibid., 6.
57 William Bonvillian, "Power Play, the DARPA Model and U.S. Energy Policy," *Holidays* (November/December, 2006), 43.
58 David Halberstam, *War in a Time of Peace: Bush, Clinton and the Generals* (New York: Simon & Schuster, 2001).
59 Peter W. Singer, "How to Be All That You Can Be: A Look at the Pentagon's Five Step Plan for Making *Iron Man* Real," *Brookings Institution* (November 17, 2009), 8–9.

4
Intended and Unanticipated Consequences

Abstract: *The process of innovation through adaption and use is impacted by multiple turns in the road, many of which result in unseen and unintended consequences. While this is important to civilian applications, it is especially important when these consequences are experienced by military organizations which are tasked with the immediate responsibility to project violence on behalf of the state. Traditional constraints within military culture include the law of war, the warrior ethic, and the relationships which have traditionally been forged between soldiers and their leaders. Emerging technologies challenge these constraints and threaten traditional military efficacy itself.*

> *Part one: What does it mean to be a warrior?*
> *Part two: What does it mean to be a civilian?*
> *Part three: Consensual risk? Soldiers, uncertain technology, and informed consent*
> *Part four: The intersection of military and civilian professional standards of care*
> *Part five: Dependence on cyber technology*

O'Meara, Richard Michael. *Governing Military Technologies in the 21st Century*. New York: Palgrave Macmillan, 2014.
DOI: 10.1057/9781137449177.0008.

> We have been given a world to live in which is inherently unpredictable. That's the bad news and the good news, all at once.[1]

The term science is derived from the Latin *scientia*, meaning "knowledge." Its project is to build and organize knowledge in the form of testable explanations and predictions about the natural world.[2] Technology, on the other hand, speaks to the usage and knowledge of tools, techniques, crafts, systems or methods of organizations. In 1937, Read Bain, a sociologist, wrote that "...technology includes all tools, machines, utensils, weapons, instruments, housing, clothing, communicating and transporting devises and the skills by which we produce and use them."[3] Cultures have been defined in terms of technology. Neil Postman, for example, classifies cultures into three types: tool-using cultures, technocracies, and technopolies.[4]

In technocracy, tools play a central role in the thought-world or the culture. Everything must give way, in some degree, to their development. The social and symbolic worlds become increasingly subject to the requirements of that development. Tools are not integrated in the culture; they attack the culture. They bid to become the culture. As a consequence, tradition, social mores, myth, politics, ritual, and religion have to fight for their lives.[5] Indeed, according to Postman, there is a deification of technology which causes a disintegration of traditional beliefs and social structure. There is an assumption, even a dedication to the idea, that technical progress is man's ultimate achievement and the place where all of man's deepest questions and puzzles can be solved. There is no downside to technological achievement since the application of scientific principles will tackle and ultimately alleviate any unintended consequence that may occur along the way.[6]

The limits of this pursuit appear to many to be nonexistent. Simon Young in his Transhumanist Manifesto notes that human beings are presently bound by a three-part genetic program reading "survive, reproduce, and self-destruct." The advent of bioscience, what he calls, Superbiology, will break this evolutionary chain. Evolution, as a result of the freedoms created by bioenhancements, will result in a stronger, more diverse species capable of controlling its own genetic makeup. Humanity, then, will not need to rely on the vagaries of nature to evolve but will define and control the process of evolution itself.[7]

Ray Kurzweil speaks to the speed of technological change as game-changing as well. History teaches that the rate of change in societies has not always been constant. Indeed, there have been ups and downs, periods of rapid change and periods of seemingly no change at all. Yet the 20th and now the 21st centuries are now experiencing a rate of change categorically different than in the past. He believes that "...we won't experience one hundred years of technological advance in the 21st century; we will witness on the order of twenty thousand years of progress (again, when measured by today's rate of progress), or about one thousand times greater than what was achieved in the 20th century."[8]

For Kurzweil, there is no methodology for governance or regulation. "Innovation," he believes, "has a way of working around the limits imposed by institutions. The advent of decentralized technology empowers the individual to bypass all kinds of restrictions, and does represent a primary means for social change to accelerate."[9]

The intended consequence of the pursuit of technology is, then, as Bernard Stiegler has stated "...the pursuit of life by means other than life."[10]

For the military, the imperatives are somewhat different. Ensconced firmly in the duties and responsibilities of the Westphalian system, the military's purpose is, and will continue to be for the foreseeable future, defense of the state against all enemies "...foreign and domestic". The United States Military Oath of Office reads as follows: "I (NAME), do solemnly swear (or affirm) that I will support and defend the Constitution of the United States against all enemies, foreign and domestic; that I will bear true faith and allegiance to the same; and that I will obey the orders of the President of the United States and the orders of officers appointed over me, according to regulations and the Uniform Code of Military Justice. So help me God."

Article 1 of the Code of Conduct requires each member of the military to recognize and commit to the following: "I am an American, fighting in the forces which guard my country and our way of life. I am prepared to give my life in their defence."[11]

Application and *use* of emerging technologies, here, are practiced for very specific and often emergent reasons. There is a tension between *experimentation*, and the failures that often accompany it (which are generally unacceptable) and *fielding* as quickly as possible the most efficient methodologies in order to complete military missions. Further, military culture is, one could argue of necessity, communal in nature; morale, discipline and, indeed, efficiency revolve around adherence and loyalty to communal values and bonds of comradeship. Keegan's review of the culture of warfare makes this point eloquently:

[T]hey are those of a world apart, a very ancient world, which exists in parallel with the everyday world but does not belong to it. Both worlds change over time and the warrior world adapts in step to the civilian. It follows it, however, at a distance. The distance can never be closed, for the culture of the warrior can never be that of civilization itself. All civilizations owe their origins to the warrior; their cultures nurture the warriors who defend them and the differences between them will make those of one very different in externals from those of another.[12]

Hence, the tendency toward "us and them" and "if it saves American lives, do it!" Yet there is a military ethic which is challenged by new technologies. Ethics is a large subject, the multiple discussions of which are outside the scope of this book. On the other hand, it is important to note that while the study of ethics, generally, deals with issues of *right* and *wrong, good* and *bad, moral* and *immoral* and the like, the concept of military ethics serves very immediate and utilitarian goals given the environments in which it is practiced. One commentator puts it this way:

> Military ethics serve as a normative code of behavior for the armed forces of a state, acting as a mechanism of definition and control within the force, between the force and its client, and between the force, its adversaries and the wider public. They have two intrinsically linked functions: a preventative function, which defines the moral and legal parameters of conduct, and a constructive function, which creates and maintains an effective and controllable force...Despite the reduction in conflict intensity, the constructive function has a remaining utility through its mediation and amelioration of the stressors engendered by the growing complexity of the operational environment.[13]

Military ethnics, then, are a method of governance. They provide explanations regarding the right of the fight (*jus ad bellum*), the right of the way we fight (*jus in bello*), and the right of the peace that is ultimately secured (jus post-bellum). They are normative rather than punitive and they are enforced not by law but rather by one's personal sense of loyalty to peers and groups of peers (the regiment, etc.) Finally, they seek to insure that soldiers self-regulate. Ethics are, as the old saying goes, how we conduct ourselves when no one is watching.

The U.S. Army spends a good deal of time teaching values, for example, and denominates *loyalty, duty, respect, selfless service, honor, integrity,* and *personal courage* as the seven core Army values which define "what being a soldier is all about."[14]

Innovation creates ethical dilemmas on a daily basis for all who get wrapped up in it, from the promoter, to the funder, to the adapter, to the user. Military officers involved in the process of procurement and development are routinely required to *think outside the box*, that is to envision needs for near and far term use and set the process of *innovation* and *adaption* in motion. Funders move between myriad possibilities as they attempt to determine which innovations are entitled to their attention and *push*; commander*s* wrestle with fielding and using technologies in order to accomplish their two responsibilities—accomplishing the mission and seeing to the safety of their soldiers; and users—those in the field—*adapt* innovation from the minute it is made available to them in multiple environments which often require split second decisions about the life and death of those around them. Further, this environment is considerably different than the one in which the ethical norms were created. Current operations are "...generally justified on moral principles and involve a multinational, joint and interagency deployment sent to intervene in an irregular, intrastate conflict occurring in an underdeveloped region and conducted under the intense glare of the media."[15]

There is a tension, then. Emergent technologies provide *fixes* for users badly in need of them; they often make the difference between life and death for the user; and they make it possible for the user to accomplish the myriad tasks required of them by civil authorities. On the other hand, their efficacy is often untested which places users in precarious situations, their unintended consequences are realized in these same environments, and their immediate uses often run well ahead of the norms and practices upon which the military institution is based. Some examples of this tension are instructive.

Part one: What does it mean to be a warrior?

At least for the foreseeable future a soldier is a human being; one who enters *the profession* with values and ethics learned at his[16] mother's knee, during his formative years in civil society, and a sense of other moral systems such as religious beliefs and so on. He is also capable of exhibiting what are generally accepted psychological traits of human beings including fear, love, anger, rage, guilt, mercy, hope, faith, generosity, courage, shame, cowardice, and so on. The warrior has traditionally been *enhanced* by training and technology to accomplish *the military function*, which, according to Samuel Huntington is performed "...by a public bureaucratized profession expert in the management of violence and responsible for the military security of the state."[17] He is also a volunteer, or at least has agreed in one form or another to enter a special class of citizens, prepared to project violence on behalf of the state and committed to the knowledge that he may be targeted by others as a result of this commitment. The warrior culture and the warrior ethic which supports it have a number of characteristics which are relevant to the definition according to Huntington. They recognize and react to the permanence, irrationality, weakness, and evil in human nature. They are concerned with the predominance of society over the individual and think in terms of hierarchy and division of function. Military culture is informed by the value of history. It recognizes the nation-state as the highest form of political organization and concedes the likelihood there will most probably be warfare between these states in one form or another for the foreseeable future. For this reason, the issue of power and state security are paramount and the need for a strong military is, therefore, self-evident. While the military is at all times training for warfare, it urges restraint by the state's leaders restricting the projection of force to activities in support of direct national interest. Consistent with these beliefs, war is considered an instrument of politics, the military is a servant the state, and military conduct is governed by a separate set of ethical proscriptions which define it as a profession. In sum, Huntington concludes "[T]he military ethic is pessimistic, collectivist, historically inclined, power-oriented, nationalistic, militaristic, pacifist, and instrumentalist in its view of the military profession. It is, in brief, realistic and conservative."[18]

Consistent with the past, the modern warrior respects actions of his peers which reflect *valor, loyalty,* and *adherence* to the military ethic, even under the direst circumstances. Because he is a realist and assumes human weakness and frailty—indeed, trains his whole life to overcome these characteristics in himself—actions which reflect these values provide *honor,* a much sought after commodity.[19] This ethic, it would appear, has two functions which are especially important given the environment in which he works. The ethic helps him differentiate between the killing he is required to do and simple murder. He is constrained to project force only in certain restricted situations. If he complies, despite the circumstance, he is deemed *honorable*; otherwise he is a thug, a base murderer, rapist, sadist, and the like. The ethic, therefore, provides constraint. Second, it can help him justify the force he has used, which provides a useful psychological benefit, contributes to morale, and personal adherence to regulation.[20] The warrior is a representative of the state for which he fights. This system of constraints inures not only to him personally and the community in which he serves, but to the state itself.

It can be argued that a system of bioenhancement, through nanotechnologies, prosthetics and/or pharmaceuticals, may well be capable of relieving the warrior of the frailties the warrior code is designed to guard against. Physical frailties can certainly be ameliorated. Further, the new soldier will no longer need to worry about fatigue, disease on the battlefield, and a whole host of other maladies which have plagued him for centuries. His *post-bellum* health concerns from psychological maladies to amputations and disfigurements can be cured as well through a vast menu of technologies. Gone will be anxieties traditionally connected with the enterprise of war, the fear, the pain, the imminence of death. Second-thoughts, guilt and shame, and pride can also be dissipated as can the need to question the projection of force in the first place. Gone too, it would appear, would be the necessity of personal achievement, the proverbial *thrill of victory and agony of defeat*. What difficulties there might be can be adjusted after the fact on an individual basis, no need for condolence or support from fellows, no identification with comrades, no concerns over valor, loyalty, and honor. One is tempted to discount these conclusions as the romantic ravings of individuals who have forgotten—or never known—the horror of the battlefield; yet they would appear to be the logical extension of the progress toward making the entire project of war pain free in order to obtain optimal efficiency, in itself a worthy goal.

Robotics, in a sense, represents merely one more type of enhancement, albeit an enhancement so great that it may take the new soldier off the battlefield completely; especially if we are able to increase the independence of robots through artificial intelligence to the point where all tasks can be completed by them.[21] At present, autonomous weapons (AW) already provide considerable capabilities for the user over and above what the warrior can accomplish. Unmanned systems are one-third the cost of manned platforms and cost two-thirds as much to operate; they reduce the kill chain (find, fix, track, target, engage, assess) from hours to minutes; they can be prepositioned thereby reducing large logistics footprints; they are not mission specific, which is to say they can be used for a wide range of missions and in a wide range of operations from conventional warfare to peace-keeping to humanitarian relief; they are persistent, that is they can remain on target for extended periods of time; and can provide *post-mortem* analysis through the use of accurate data; and they are capable of precision strikes.

The unanticipated consequence, then, involves the question—is it a good thing to make war a painless exercise akin to a video game or a week at a dude ranch? Is there some value to the warrior's code which is lost when the personal stakes are no longer high? Can we enhance the biological body to a point which is inconsistent with the definition of what it means to be human? And does this make any difference?

In addition, issues of inequality are raised. Does every new soldier get the benefits of enhancement or only those deemed worthy through some sort of medical and means testing? Who gets to be an iron man, enhanced to the point of complete protection while others must continue to endure the vagaries of the battlefield? What of the intersection between these new iron men and civil society? If all new soldiers can be equally smart, equally brave, equally fit, and equally competent, what purpose will exist for the hierarchical nature of the institution, with its paternalism and emphasis on leadership? On what basis will authority rest?

Finally, for civil authorities who must ponder their use of these new soldiers, will it be easier to start wars and continue the projection of force in the knowledge that there will be no body bags, minimal suffering, and long-term consequences for the body politic? Will this phenomenon continue to widen the gap between those who order the projection of force and those who accomplish it?[22]

Part two: What does it mean to be a civilian?

A second question involves the status of the myriad individuals who project force on behalf of the state but have not agreed to their classification as warriors.[23] This discussion, of course, requires better definitions of "civilian" than are found in humanitarian law, which presupposes that those individuals are not geographically located on a battlefield, or, if on a battlefield, are not holding themselves out as combatants through the use of uniforms, hierarchies of command and other indicia of membership in military organizations. There are presently more than 700,000 Department of Defense employees and well over 100,000 civilian contractors who have served in Iraq and Afghanistan, none of whom have presumably agreed to take up arms against all enemies *foreign and domestic* and most of whom would be surprised to find out that their employment status qualifies them for special handling when it comes to targeting.[24]

Traditionally, of course, International Humanitarian Law (IHL) has a decided repugnance for the targeting of civilians who are *hors de combat*, for whatever reason.[25] The point is that as stand-off weaponry becomes more sophisticated through the use of information technology, biotechnology, and space technology, and so on, it becomes possible to project violence or aid in the projection of violence from civilian centers far removed from the traditional geography of the battlefield. Already, uniformed *pilots* in the United States position *predator* aircraft which kill and maim targets of opportunity in Afghanistan and Pakistan and elsewhere, causing collateral damage, like death to unarmed civilians.[26] Is the civilian technician who maintains the equipment for the pilot in Nevada a legitimate target; is the civilian secretary who makes up the manifest and otherwise enhances the ability of the pilot to accomplish his mission a legitimate target? What responsibility does the state have for the protection of these individuals over and above what it owes the average citizen as a result of their status? Are they entitled, for example, to jump to the head of the line when vaccines are handed out? Do they have access to specialized bunkers in the event of nuclear warfare? Should their conduct be subject to specialized laws such as the Uniform Code of Military Justice? Is desertion from their place of employment during an emergency, for example, of such importance that it should be punishable by death as is the case for their *co-workers* in uniform?

Part three: Consensual risk? Soldiers, uncertain technology and informed consent

Conducting experiments on military personnel has a long and often sad history. Traditions of obedience, group requirements versus individual rights, and the emergent nature of new and dangerous threats coalesce to create an environment in which experimentation or *testing* had been justified as immediately necessary in order to accomplish the mission. Medical experimentation, for example, often requires large study groups with homogenous populations of healthy individuals who can be studied over relatively long periods of time. Indeed, there are very few other organizations where these conditions exist.[27] Two examples of state-run, yet non-military, medical experiments are the U.S. medical research in Bilibid prison located in the American-occupied Philippines on prisoners to determine the efficacy of a cholera vaccine and the 1932 Public Health Service Tuskegee Syphilis Study. Here more than 400 African-American men, suffering from syphilis were actively misled regarding their participation in the study and denied the benefits of penicillin. Multiple deaths occurred in these experiments and the manner in which they were conducted has led to a tradition of distrust of the medical community and the government which sponsored them. Further, military personnel have been subjected to multiple technologies without studied scientific determinations regarding their short-term or long-term effects. Between 1954 and 1973, some 2,300 Seventh Day Adventists served as conscientious objector *volunteers* in 137 protocols in defensive biological weapons testing. These experiments were directed at developing and "...testing vaccines and therapeutic drugs against Q fever, tularemia, various viral encephalitis, Rift Valley fever, sand fly fever, and plague."[28] The Cold War produced numerous experiments on soldiers ranging from open air tests of radiological and bacterial materials[29] to LSD testing in order to determine the efficacy of the drug as a truth serum (also used in interrogations).[30] In Vietnam, military personnel were subjected to Agent Orange as part of a fairly substantial deforestation program which resulted in multiple cancers and birth defects.[31] And there is a good deal of evidence that military personnel and their families have been living with unknown degraded environmental hazards as well.[32]

Ironically, the military has developed a fairly robust set of rules and regulations regarding medical testing and experimentation over the years which can be said to rival and in some cases best its civilian counterparts.[33] They reflect the U.S. military's reaction to Nazi medical experimentation during World War II. International Law speaks to these concerns and military medical proscriptions in the form of regulation and practice mirror the tension between obtaining useful information that will aid in the accomplishment of stated missions and perceived and actual abuse which can result. The Nuremburg Principles contain multiple constraints. The first is of particular relevance here.

Directives for human experimentation

Nuremberg code

1 The voluntary consent of the human subject is absolutely essential. This means that the person involved should have legal capacity to give consent: should be so situated as to be able to exercise free power of choice, without the intervention of any element of force, fraud, disease, duress, over-reaching, or other ulterior form of constraint or coercion, and should have sufficient knowledge and comprehension of the elements of the subject matter involved as to enable him to make an understanding and enlightened decision.

This latter element requires that before the acceptance of an affirmative decision by the experimental subject there should be made known to him the nature, duration, and purpose of the experiment, the method and means by which it is to be conducted; all inconveniences and hazards reasonably to be expected; and the effects upon his health or person which may possibly come from his participation in the experiment. The duty and responsibility for ascertaining the quality of the consent rests upon each individual who initiates, directs or engages in the experiment. It is a personal duty and responsibility which may not be delegated to another with impunity.[34]

Mindful of these rules, military experimentation requires that the standard Institutional Review Board (IRB) proceedings mandated for all biomedical testing in the United States contains a separate level of review above that, and must always include *a therapeutic component*.[35]

This regulatory system of protections is bottomed on rigorous informed consent requirements which, it has been argued, make human trauma and emergency research almost impossible.[36] There is, therefore, a tension between the need for experimentation and the responsible protection of individuals involved in the experiments.

Further, it is questionable whether in a culture which instructs—indeed demands obedience, if not reverence for authority—young men on and off the battlefield are capable of exercising independent judgment regarding sophisticated issues which are the subject of scientific experimentation. Soldiers constitute, perhaps, the ultimate vulnerable population, given this adherence to orders, potential for coercion by superiors, and the environment in which they make their decisions (battlefield, communal). Is it possible for a soldier to be protected from the formal or informal coercion of a superior and at the same time commit to that superior's right and obligation to order him into battle?[37]

Military commanders, as a matter of ethical obligation, assume responsibility for the safety of their men and increasingly their dependents. They weigh this responsibility against accomplishment of the missions set before them and they are prepared to risk that safety in order to accomplish that mission or to insure the welfare of the whole over the individual. Soldiers understand this. It is part of the unwritten contract between soldier and commander. Thus, soldiers are routinely put in harm's way, tolerate unhealthy and dangerous environments, work with unsafe technologies and otherwise risk life and limb in the belief that commanders have reasons for the decisions they make. Commanders themselves, however, rely on their superiors to provide them with technologies that have a fair degree of efficacy before they are fielded. It is not in the commanders' lexicon to tell a soldier that the commander is not responsible for the failure of a technology or that he really has no opinion regarding the subject of a soldier's consent to an experiment. There is a phrase in the culture of leadership which is instructive—*the superior is responsible for everything his soldiers do or fail to do*. It can be argued that there is simply no place in the relationship for independent decisions by soldiers, especially about important matters. When technology is fielded and doesn't work, when it causes severe and unintended consequences, when it, not the enemy, threatens the soldier there is a delegitimization of authority. These conditions strike at the heart of military organizations and make them inefficient and incapable of performing the difficult tasks set before them.[38]

The intersection between experimentation, perceived failure of technology, and commander's authority is illustrated in the anthrax scare of the 1990s. Commanders have the legal right and responsibility to require service members to undergo various medical procedures, including treatments for injuries, psychological counseling, vaccinations, and medical examination. These medical procedures are ordered both for the good of the service generally—to insure efficiency on the battlefield—and in compliance with the superior's ethical obligation to provide for the safety of the soldier. What happens when soldiers, empowered to rely on personal choice (the right to informed consent regarding experimentation and to denial of illegal orders), perceive that a specific procedure is potentially unsafe or not efficacious? Such was the case with the anthrax vaccination, ordered for all soldiers in 1998. Concerned over issues of sterility and other side effects, soldiers began refusing to comply with orders to take the vaccine, thus rendering them non-deployable in the eyes of their superiors, akin to refusal to wear helmets and flack gear, or fire a weapon. A history of the innovation of the vaccine was not helpful, given the fact it had been approved by the FDA primarily for anthrax sustained as a result of personal contact with infection rather than inhalation. Also of interest were the facts that surrounded the development and manufacture of the drug were readily available to the troops on line. Indeed, when the FDA approved the vaccine as safe, efficacious, and not misbranded it noted that the "...anthrax vaccine poses no serious special problems other than the fact that its efficacy against inhalation anthrax is not well documented." It was, precisely, inhalation anthrax that was proposed as the justification for the use, however. Further, there had been problems with regard to its manufacture in the one facility licensed to provide the substance. And, finally, the Institute of Medicine, while confirming that no long-term effects were known to exist, also noted "...that research is currently insufficient to allow us to draw long-term conclusions." When soldiers refused to take the medication, a multitude of disciplinary actions were taken, forced resignations, and lawsuits in federal courts. Allegations that these vaccinations have caused injection site hypersensitivity, Guillain-Barre syndrome, multiple sclerosis, anaphylaxis, and Gulf War syndrome have caused soldiers to question the good faith intentions of their superiors.[39] As Susan Leder has concluded: "[L]awsuits brought by veterans of biological, chemical, and atomic warfare studies continue to wend their way through the courts. The lawsuits permit a financial accounting of loss of life, liberty, and mental distress. They do not take into account the corrosion of trust in American researchers and the American government. Even more disturbing is the fear that these things could happen again unless adequate safeguards remain in effect and the lessons of the past are learned."[40] How can the military commit its forces in a timely manner, if soldiers elect not to accomplish the myriad tasks necessary to certify themselves and their units capable to deploy?

Part four: The intersection of military and civilian professional standards of care

There is a separate set of concerns arising from the intersection between the use of emerging military technologies and practices and civilian standards of care. Professionals in the military who *innovate* and *adapt* these technologies and practices are licensed by their respective disciplines, engineers, medical professionals, lawyers, psychologists and so on. Indeed, their state licensure in good standing is a condition for their continued service in the military. What happens when a particular military practice is determined to be in violation of a particular state or national code of ethics? This issue has arisen in conjunction with military psychology and medical practice and its use during interrogation of suspected terrorists since 9/11. It has been fairly well documented that medical personnel, doctors, medics and psychologists were routinely involved in various ways in the interrogation techniques after 9/11 which have formed the basis of the debate over interrogation/torture. Sharing of information from therapeutic records with interrogators, advising interrogators regarding psychological weakness, useful techniques and practices, actually doing the interrogations, keeping records of interrogation experience (experimentation?) and otherwise being insinuated into the entire process of detaining and exploiting prisoners for the purpose of obtaining information—these and other practices which it has been argued by many constituted *torture,* have been fairly commonplace.[41] They are now roundly condemned by fellow professionals at the state and national level.[42]

A review of other professional codes of ethics will discover similar pronouncements regarding *right* professional conduct and adherence thereto. Anthropologists, for example, who have been contracted by the U.S. Government to work with military personnel in Afghanistan as part of the Human Terrain System Program (HTS), have been roundly criticized as well. In a letter to Congress written in January 2010, the Network of Concerned Anthropologist (NCA) took the position that:

> HTS is unethical for anthropologists and other social scientists. In 2007, the Executive Board of the AAA [American Anthropologists Association] determined HTS to be "an unacceptable application of anthropological expertise". Last December [2009], the AAA commission found that HTS "can no longer be considered a legitimate professional exercise of anthropology" given the incompatibility of HTS with disciplinary ethics and practice. Like medical doctors, anthropologists are ethically bound to do no harm. Supporting counterinsurgency operations clearly violates this code. Moreover, the HTS program violates scientific and federal research standards mandating informed consent by research subjects.[43]

These codes were generally created in contemplation of their respective civilian practices and do not presume to incorporate the exigencies of the military environment and culture. Will it be necessary in the future to revamp licensing practices to create separate ethical codes for military professionals? Should civilian professional review boards continue to judge the professional activities of military professionals? Is there a difference between the ethical duties of military professionals and their civilian counterparts?

Part five: The dependence on cyber technology

Secretary of Defense, Donald Rumsfeld, when asked by soldiers in Iraq why the Department of Defense was not providing armored vehicles in response to increased insurgency IED attacks famously answered, "[Y]ou go to war with the Army you have, not the Army you might want or wish to have."[44] This answer, while criticized at the time, reflects a truth about military force projection. Rarely is the threat which requires the projection properly anticipated or planned for. General J.N. Mathis of the U.S. Joint Forces Command, the military organization primarily responsible for looking into the future, discovering the nature of threats, and preparing the military for them, acknowledged this reality. Speaking about the future of warfare in the 21st century, he notes that inevitably "... [W]e will find ourselves caught off guard by changes in the political, economic, technological, strategic, and operational environments. We will find ourselves surprised by the creativity and capability of our adversaries. Our goal is not to eliminate surprise—that is impossible. Our goal is, by a careful consideration of the future, to suggest the attributes of a joint force capable of adjusting with minimum difficulty when the surprise inevitably comes."[45]

Cyberspace permeates nearly every aspect of societies from personal computers and cell phones to networked transportation and inventory systems. Aside from its obvious advantages, cyberspace carries with it a countervailing characteristic. There is simply no aspect of modern daily life that is not dependent on our ability to access in a rapid manner the multiple networks which bind us all together. Even as we depend on highways and oceans, the internet provides a separate set of trails upon which to journey. We conduct all manner of business, purchase goods, entertain ourselves, find partners, express and define ourselves, and run our basic utilities in cyberspace. Our ability to use cyberspace amplifies every aspect of our national power. Indeed the ability to maneuver and defend ourselves in cyberspace is an emerging definition of power itself. Further, like spaces on land and sea, cyberspace is not benign. America's adversaries are also capable of cyberspace use and therefore have the ability to attack, degrade, and disrupt communications and the flow of information. Dependence on this most valuable technology carries with it multiple vulnerabilities which impact the military's ability to operate in traditional ways.[46]

William Lynn III, former U.S. Deputy Secretary of Defense, acknowledges this dependence as well. Simply put:

> [I]nformation Technology enables almost everything the U.S. military does: logistical support and global command and control of forces, real-time provision of intelligence, and remote operations. Every one of these functions depends heavily on the military's global communications backbone, which consists of 15,000 networks and seven million computing devices across hundreds of installations in dozens of countries. More than 90,000 people work full time to maintain it.[47]

There are reasons for this dependence. Clearly the ability to range free throughout the globe with the information and intelligence necessary to identify threats and better control the battlefield is an enhancement of monumental proportions. Information acts as what is commonly referred to as a *force multiplier*, that is, that as a result of this technology, less men, in shorter periods of time, with smaller logistical trails and essentially less baggage of all kinds can respond to threats with more force thereby achieving better results. As with other technological enhancements, information technology has political ramifications as well including the projection of force on the battlefield with less loss of life, military and civilian.

What happens, however, when critical services are degraded or completely neutralized as a result of cyber attacks? As Wesley K. Clark notes, "[a]n electronic attack is extremely cheap, is very fast, can be carried out anonymously, and can disrupt or deny critical services precisely at the moment of maximum peril."[48]

Nor are these vulnerabilities and dependencies restricted to the ability of the military to *fight and win America's wars*. Significantly, it is recognized that the definitions of battlefield and the nature of warfare itself are changing. Michael L. Gross notes that traditional humanitarian practices between states seeking political accommodations in war are giving way, especially where the mission is determined to be the ouster of a regime, the elimination of a terrorist group or the eradication, in the cheapest way possible, of a superpower's ability to amass and practice power on the international stage.[49] Here, all bets are seemingly off, all constraints abandoned, especially those involving the targets of attacks. Thus civilians, economies, infrastructures, indeed, governmental legitimacy, are fair game. Cyberwarfare is especially relevant here. The National Strategy to Secure Cyberspace recognizes this dilemma:

> By exploiting vulnerabilities in our cyber systems, an organized attack may endanger the security of our Nation's critical infrastructures. The vulnerabilities that most threaten cyberspace occur in the information assets of critical infrastructure enterprises themselves and their external supporting structures, such as the mechanisms of the Internet. Lesser-secured sites on the interconnected network of networks also present potentially significant exposures to cyber attacks. Vulnerabilities result from weakness in technology and because of improper implementation and oversight of technological products.[50]

It can be argued that it is not merely the vulnerability to attack that causes the problem but, perhaps more importantly, the dependence on cyber technologies which bears watching.

Anecdotally, the author is reminded of a situation which occurred during Operation Desert Storm in 1990. Required to mobilize multiple units of men on short notice in the US Army Reserve, the author maintained and utilized a computer network and database which reported the strength and capabilities (training assessments, logistical status, etc.) of each unit. Types of units were identified by the Pentagon and requirements were communicated to the author who would see to the deployment of the units. Within a month, the system broke down completely. There was a wide ranging fear that the information recorded was inaccurate or poorly communicated. Further, there was a distrust of the recorders of the information. The network was abandoned and units began to be called up based upon late-night phone calls from Pentagon planners to the author who was required to render opinions on the spot regarding the efficacy of each unit. This was of particular importance just prior to the actual invasion of Kuwait when the requirement for medical personnel, hospital equipment and supplies were identified in anticipation of considerable casualties. Cyberattacks are, as has been seen, capable of striking at civilian infrastructure as easily as military infrastructure. Query, if the telephone lines had been degraded as well? The unanticipated consequence here, then, is the inability to project force on a timely basis when the technology fails. As Wesley Clark notes, "[W]hen it comes to cybersecurity, Washington faces an uphill battle." And as a recent Center for Strategic and International Studies report put it, "It is a battle we are losing."[51]

Note

1. Tim Healy, "The Unanticipated Consequences of Technology," Markkuyla Center for Applied Ethics, Santa Clara University, 2010, 11, retrieved at http://www.scu.edu/ethics/publications/submitted/healy/consequences.html, October 26, 2010.
2. *Merriam-Webster Dictionary*, retrieved at http://www.m-w.com/disctionary/science, October 26, 2010.
3. Read Bain, "Technology and State Government," *American Sociological Review* 2 (December 1937), 860.
4. Neil Postman. Technopoly: *The Surrender of Culture to Technology* (New York: Vintage Books), 22–48.
5. Ibid.
6. Ibid.
7. Simon Young, *Designer Evolution, a Transhumanist Manifesto* (New York: Prometheus Books, 2006), 38.
8. Ray Kurzweil, *The Singularity Is Near: When Humans Transcend Biology* (New York: Penguin Books, 2006), 11.
9. Ibid., 473.
10. Bernard Steigler. *Technics and Time, 1: The Fault of Epimentheus* (Stanford: Stanford University Press, 1998), 17.
11. Article 1, United States Code of Conduct, retrieved at http://usmilitary:about.com/od/justicelawlegislation/a/codeofconduct1.htm, November 26, 2009.
12. Keegan, *History of Warfare*, xvi.
13. Mathew William Rout, "To Define & Control: the Utility of Military Ethics in the New Zealand Army's Contemporary Operations Environment," Master's Thesis, University of Canterbury, 2009, retrieved at http://cantebury.ac.nz/handle.10092/3048, Novemebr 26, 2009.
14. Department of the Army, "Soldier Life, Living the Army Values," retrieved at http://www.goarmy.com/life/living_the_army_values.jsp, November 27, 2009.
15. Rout, "To Define and Control."

16 One must remember that the definition of the contemporary warrior is no longer gender specific, especially in the United States. Some 20 percent of the military force is comprised of women, including 14.2 percent active force, 24.1 percent Reserve force, and 14.1 percent National Guard, who have proven capable of accomplishing most, if not all, primary skills of soldiering. *Women in Military Service For America Memorial Foundation, Inc.* hg.womensmemorial.org. The use of the male gender here, then, is done for simplicity's sake *only* and should not be construed as an attempt by the author to enter in any way the discussion regarding the efficacy of female soldiers, a discussion which has been contentious and often poorly articulated.

17 Samuel P. Huntington, *The Soldier and the State: The Theory and Politics of Civil-Military Relations* (Cambridge, MA: The Belknap Press of Harvard University Press, 1957), 61.

18 Ibid., 79. Regarding war itself, Huntington continues, "[H]e is afraid of war. He wants to prepare for war. But he is never ready to fight a war." 69.

19 Walter Lippmann is often quoted in this regard, "[A] man has honor if he holds himself to an ideal of conduct though it is inconvenient, unprofitable, or dangerous to do so." Walter Lippman, retrieved at http://thinkexist.com/common/print.asp?id=226812&quote=a_man_has_honor_if_he_hold..., November 30, 2009.

20 Shannon French, "The Warrior's Code, 2001." "Before we call any collection of belligerents a culture of warriors, we should first ask why they fight, how they fight, what brings them honor, and what brings them shame," retrieved at http://www.au.af.mil/au/awc/awcgate/jscope/french.htm, November 26, 2009; See also, Shannon French, *The Code of the Warrior: Exploring Warrior Values Past and Present* (Lanham, MD: Rowman & Littlefield Pub., Inc, 2005); John McCain, *Faith of My Fathers, A Family Memoir* (New York: Harper Perennial, 2000).

21 Guetlein, "Lethal Autonomous Weapons," 4–5.

22 This has, arguably, already occurred to a large extent. As Secretary of State Madeline Albright asked Colin Powell in the 1990s, "What's the point of having this superb military you're always talking about if we can't use it?" Sheldon Richman, "Clinton's Quagmire," *Freedom Daily*, July 1999, retrieved at http://www.fff.org/freedom/0799c.asp, November 26, 2009.

23 Id.

24 Department of Defense, "It Takes More Than Soldiers to Protect America," retrieved at http://www.go-defense.com/, November 20, 2009; Renae Merle, "Census Counts, 100,000 Contractors in Iraq," *The Washington Post*, December 5, 2006, retrieved at http://www.washingtonpost.com/wpdyn/content/article/2006/12/04/AR2006120401311.htm., November 29, 2009.

25 Humanitarian law is that international law comprised of a set of rules which seek to limit the effect of armed conflict. Primary conventions

include the Geneva Conventions of 1949, supplemented by the Additional Protocols of 1977 relating to the protection of victims of armed conflicts; the 1954 Convention for the Protection of Cultural Property in the event of armed Conflict and additional protocols; the 1972 Biological Weapons Convention; the 1980 Conventional Weapons Conventions and its five protocols; the 1997 Ottawa Convention on anti-personnel mines; and the 2000 Optional Protocol to the Convention on the Rights of the Child on the involvement of children in armed conflict. International Committee of the Red Cross, "What is International Humanitarian Law?" Advisory Service, 2004, retrieved at http://www.icrc.org/Web/Eng/siteengo.nsf/ htmlall/section_ihl, November 25, 2009.

26 Singer, *Wired for War,* 326–28. Strikes in Pakistan began in 2004 and have increased in number and lethality ever since; 2004 (1), 2005 (1), 2006 (3), 2007 (5), 2008 (36) and 2009 through September (42). Civilian casualties, to the extent they can be determined with any accuracy appear relatively low as well: 20 civilian to 122 Taliban/Al Qaida (enemy) in 2006; 0 civilian to 73 enemy in 2007; 31 civilian to 286 enemy in 2008 and 43 civilian to 404 enemy in 2009. Bill Roggio, Alexander Mayer, "Analysis: A look at US airstrikes in Pakistan through September 2009," *The Long War Journal*, October 1, 2009, retrieved at http://www.longwarjournal.org/archives/2009/10/ analysis_us_airstrik.php, November 25, 2009.

27 John McManus, Sumeru G. Mehta, Arnette R. McClinton, Robert A. DeLorenzo, Toney W. Baskin, "Informed Consent and Ethical Issues in Military Medical Research." *U.S. Army Academy of Emergency Medicine,* vol. 12, no. 11 (November 2005), 1121.

28 Ibid.

29 American Patriot Friends Network, "Secret U.S. Human Biological Experimentation," May 2004, retrieved at http://www.apfn.org/apfn/ experiment.htm, November 25, 2009; *see also* Howard L. Rosenberg, *Atomic Soldiers: American Victims of Nuclear Experiments* (Boston: Beacon Press, 1980).

30 *U.S. v Stanley, 483 U.S. 669 (1987),* retrieved at http://caselaw.lp.findlaw.com/ scripts/getcase.pl?court=US&vol=483&invol=669, November 27, 2009.

31 United States Department of Veterans' Affairs, "Agent Orange: Diseases Associated with Agent Orange Exposure," *Office of Public Health and Environmental Hazards,* 2009, retrieved at http://www.publichealth.va.gov/ exposures/agentorange/diseases.asp, November 27, 2009.

32 U.S. Medicine, "Legislators Express Concern Regarding Environmental Hazards" (November 17, 2009), retrieved at http://www.usmedicine.com/articles/ Legislators-Express-Concern-Regarding-Environmen.asp, November 27, 2009.

33 McManus, "Informed Consent and Ethical Issues in Military Medical Research," 1124.

34 George J. Annas, Michael. A. Grodin, *The Nazi Doctors and the Nuremberg Code: Human Rights in Human Experimentation* (Oxford: Oxford University Press, 1992), reprinted from *Trials of War Criminals Before the Nuremberg Military Tribunals Under Control Council Law No. 10.* U.S. Government Printing Office: Washington D.C., 1949, 181–82.
35 McManus, "Informed Consent," 1122–23.; see generally, Barry Bozeman, "Understanding Bureaucracy in Health Science Ethics," 1154–55, for a discussion of Institutional Review Boards.
36 McManus, "Informed Consent," 1122.
37 Jessica Wolfendale, Steve Clarke, "Paternalism, Consent, and the Use of Experimental Drugs in the Military." *Journal of Medicine & Philosophy*, vol. 33, issue 4 (August 2008), 337–45.
38 Scranton, "The Challenge of Technological Uncertainty."
39 "Informed Consent in the Military: Fighting a Losing Battle Against the Anthrax Vaccine." *American Journal of Law & Medicine* (June 22, 2002), retrieved at http://goliath.ecnext.com/coms2/gi_0199-1833062/Informed-consent-in-the-military.html, November 25, 2009; MaManus, "Informed Consent," 1124.
40 Susan E. Lederer, "Chpt 17, The Cold War and Beyond: Covert and Deceptive American Medical Experimentation," in *Military Medical Ethics,* vol. 2. Boden Institute, Office of the Surgeon General, Washington, DC (2003).
41 See generally Senate Armed Services Committee Inquiry Into the Treatment of Detainees in U.S. Custody; Committee of the Red Cross Report, retrieved at http://www.nybooks.com/ircr-report.pdf November 25, 2009; Physicians for Human Rights, White Paper, "Broken Laws, Broken Lives" (November 28, 2007); Sheri Fink, "U.S. Medical Personnel and Interrogations: What Do We Know? What Don't We Know?" *ProPublica* (April 9, 2009); Steven Miles, *Oath Betrayed: Torture, Medical Complicity and the War on Terror.* 2006 and "Military Medicine and Human Rights," *The Lancet,* v. 364, Issue 9448 (November 20, 2004), 1851–52. All the above and many others have taken the position that psychologists complicit in the interrogation activities generally are in violation of the APA's ethical code: "Psychologists strive to benefit those with whom they work and take care to do no harm," cited in Stephen Soldz, "Ending the Psychological Mind Games on Detainees," *Op-Ed Boston Globe*, August, 14, 2008, retrieved at http://brokenlives.info/?tag=psychologists, November 30, 2009; but see Michael L. Gross, *Bioethics and Armed Conflict, Moral Dilemmas of Medicine and War* (Cambridge, MA: MIT Press, 2006), "the contemporary dilemma of torture and ill-treatment sets lives of some against the self-esteem of others... If doctors remain convinced that interrogational torture could save more lives than other forms of interrogations, avoids unnecessary harm and only targets those who have forfeited their right to

self-esteem, they may consider providing facilitating medical care during an interrogation," 220.
42 Jonathan Hutson, "After Senate Report, Psychologists Who Tortured Must Be Held Accountable," *Physicians for Human Rights*, April 21, 2009. Colombia University, *Ethics Abandoned*.
43 Network of Concerned Anthropologists, "Anthropologists Statement on the Human Terrain System Program (January 28, 2010)," retrieved at http:savageminds.org/2010/01/28concerned-anthropoligists-letter-to-Washington/, October 28, 2010.
44 NBC, msnbc.com and news services, "Bush: Soldiers' Equipment Gripes Heard" (December 9, 2004), retrieved at http://www.msnbc.msn.com/id/66/6676765/, October 28, 2010.
45 J.N. Mathis, *The JOE 2010, Joint Operating Environment* (US Joint Forces Command: February 18, 2010), retrieved at www.jfcom.mil, October 20, 2010.
46 Ibid.
47 William Lynn III, "Defending a New Domain, The Pentagon's Cyberstrategy," *Foreign Affairs*, vol. 89, no. 5 (September/October 2010), 97.
48 Clark et. al., "Securing the Information Highway," 2.
49 Michael L. Gross, *Moral Dilemmas of Modern War, Torture, Assassination, and Blackmail in an Age of Asymmetric Conflict* (Cambridge: Cambridge University Press, 2010), 2–3.
50 The White House, *The National Strategy to Secure Cyberspace* (Washington, DC: The White House, February 2003), xi.
51 Clark et. al., "Securing the Information Highway," 2.

5
Contemporary Governance and Architecture

Abstract: *Governance regarding emerging military technologies is, at best, a patchwork of treaties, norms, ethical constraints, practices, and procedures which are being adapted to meet the challenges which present themselves as innovations are designed and adaptions occur. A system of governance simply does not exist nor are there specific organizations dedicated to the analysis of ethical and legal concerns as they present themselves. Some traditional norms such as IHL and the Nuremberg Principles are being adopted to regulate the use of technologies and these will form the basis for policy as well as legal and ethical constraints in the future.*

> *Part one: Norms and ethical considerations*
> *Part two: International Humanitarian Law (IHL)*

O'Meara, Richard Michael. *Governing Military Technologies in the 21st Century.* New York: Palgrave Macmillan, 2014. DOI: 10.1057/9781137449177.0009.

Contemporary Governance and Architecture   67

At present, there are no laws or treaties specifically pertaining to restrictions or governance of military robots, unmanned platforms, or other technologies currently under consideration within the purview of this book. Instead, aspects of these new military technologies are covered piecemeal by a patchwork of legislation pertaining to projection of force under IHL; treaties or conventions pertaining to specific technologies and practices; international humanitarian law; and interpretations of existing principles of the Law of Armed Conflict (LOAC).[1]

There are, for example, multiple conventions in international law which purport to deal with specific technologies and practices, such as agreements pertaining to biological weapons,[2] chemical weapons,[3] certain types of ammunition,[4] the hostile use of environmental modification,[5] land mines,[6] incendiary weapons,[7] blinding laser weapons,[8] and numerous others.[9] The United States is not a party to all of these conventions, and to the extent their requirements do not rise to the level of customary international law, the United States is not specifically bound by them. On the other hand, the United States has taken considerable interest in the articulation of standards which purport to regulate conduct generally on the battlefield, including how weapons are used.

Part one: Norms and ethical considerations

There are, however, a variety of other potential existing constraints found in military doctrines, professional ethical codes, and public "watchdog" activities (as well as in international law) that might pertain to the present governance dilemma regarding emerging military technologies.. These constraints, generally, were created to address a variety of issues which are not wholly consistent with or applicable to the challenges created by the development and use of emerging military technologies for military and security purposes. Yet, their existence does provide architecture upon which to build a system of governance regarding the military use of military technologies on the battlefield.

It bears noting that governance systems that are successful in obtaining compliance with a particular policy, rule, or directive share a number of important characteristics. Successful systems of "good governance" involve *clearly defined and articulated* expectations: that is, they identify the precise problems to be solved, changes to be made, or goals to be sought through governance in straightforward terms. The solutions proposed to these problems, moreover, are *realistic*: that is, they do not attempt to articulate ideal norms of what *ought* to be, but rather provide feasible norms describing what *can*, in fact, be accomplished, under existing political, cultural and legal constraints. Successful systems of governance, moreover, are *holistic and inclusive*, in the sense that all stakeholders are identified and involved in some fashion in making the rules. Finally, they issue rules or principles that are *subject to assessment*: that is, the results are capable of measurement and evaluation of effectiveness, in a manner that allows for subsequent amendment and improvement of the requirements when appropriate.[10]

If these principles of good governance are not adhered to, expectations and pronouncements often go unheeded. In light of these canons of best practice for good governance, it can be argued that the goal of technological innovation governance should be to insure that all technological innovation is accomplished within the framework of a culture that respects the long-term effects of such work, while considering, insofar as possible, the likely ramifications of the proposed innovation and development. Appropriate governance should also insure that future end-users or consumers of the specified technological innovations are aware of those ramifications, ideally in the design phase, but at the very least, well before development or application of the innovations in question. All this should be accomplished, moreover, without placing too heavy of a legislative hand on, nor otherwise discouraging, the creative and competitive energies that generate much-needed innovation.

Measured against the foregoing standards, contemporary governance architecture regarding the innovation and use of emerging military technologies would appear wholly inadequate to the task. And yet, there is considerable professional, national, and international infrastructure upon which to hang a regime of articulated goals and proscriptions.

At the professional level, for example, there are multiple codes for ethical guidance regarding both best practices and limits on acceptable professional practice for a wide range of academic and professional disciplines. These codes of ethical conduct generally reflect the "do no harm" proscription specifically highlighted in the medical profession's Hippocratic Oath which is at odds with military requirements to, in fact, "do harm" to adversaries in a wide range of situations. On the other hand, these ethical codes might conceivably find themselves applied to innovation in the field of robotics, especially for participants from professions such as engineering, computer science, biology, medicine, law, and psychology. As a general rule, these ethical codes or guidelines for professional practice are grounded in the traditional responsibilities of their individual professions, and do not contemplate the challenges which can be said to presently exist for innovation generally. Professions, for example, are often regulated at the state level based upon varying degrees of oversight by private organizations and societies. Those codes speak primarily to issues of the professional's relationship and responsibilities toward clients and customers, as well as toward likely competitors, and likewise address important moral and legal issues such as privacy, intellectual property, and education, but often lack any concrete obligations relating to broader social responsibilities for technology development. The American Psychological Association, for example, does speak to "...the welfare and protection of the individuals and groups with whom psychologists work and the education of members, students, and the public regarding ethical standards of discipline." They seek to "...minimize harm where it is foreseeable and unavoidable."[11] On the other hand, when the standards are inconsistent with a requirement in law, regulations or other governing legal authority, the ethics code permits the psychologist to yield (e.g. to a government regulation or order).[12] The American Medical Association provides eleven principles which "...define the essentials of honorable behavior for the physician." Interestingly, the principles do not contain the traditional *do no harm* proscription but do require the provision of competent medical care "...with compassion and respect for human dignity and rights." In most cases, except emergencies, physicians retain the right to "...choose whom to serve, with whom to associate, and the environment in which to provide medical care" (Principle 1).[13] The American Psychiatric Association requires their members to follow the ethical prescripts of their medical colleagues. The American Society of Civil Engineers requires engineers to "...hold paramount the safety, health, and welfare of the public" and to "...strive to comply with the principles of sustainable development in the performance of their professional duties."[14] A private professional association, ISACA, which purports to serve "IT governance professionals," requires their members only to "...support the implementation of, and encourage compliance with, appropriate standards, procedures and controls for information systems."[15] There is even a code of ethics for robots being proposed by the Republic of South Korea, although the terms of the code have yet to be fleshed out. The main focus of the charter deals with social problems such as human control over robots and humans becoming addicted to robot interaction (robots as sex toys, etc.) The document will deal with legal issues, such as the protection of data acquired by robots and establishing clear identification and traceability of the machines.[16]

On occasion, but rarely, these internal ethical codes also appear to contemplate the future contexts in which professionals will have to operate. For example, a "Pledge of Ethical Conduct" printed in the commencement program for the College of Engineering at the University of California, Berkley in May 1998, reads:

> I promise to work for a BETTER WORLD where science and technology are used in socially responsible ways. I will not use my EDUCATION for any purpose intended to harm human beings or the environment. Throughout my career, I will consider the ETHICAL implications of my work before I take ACTION. While the demands placed upon me may be great, I sign this declaration because I recognize that INDIVIDUAL RESPONSIBILITY is the first step on the path to PEACE.[17]

These internal professional codes and norms are complemented by a host of non-governmental organizations (NGOs) which contribute to the transparency of innovation programs, especially those performed on behalf of the State. The goals and agendas of these organizations are as varied as their names but their methodologies generally help to educate the end-user or consumer about what is being developed and what the future may portend. Such NGOs often succeed in establishing a record of evidence and impact regarding a particular thread of innovation, and placing this evidence before the public and state funders (legislatures, policy-makers, and appropriate government agencies) and providing news media with the expertise to report on the likely ramifications of proposed technological innovations. One example is the International POPs Elimination Network (IPEN), "a unique global network of people and public interest organizations" that "share a common commitment to achieve a toxic free future. IPEN is composed of over 700 public interest health and environmental organizations from more than 80 countries."[18] Other organizations have taken specific positions regarding the ethical behavior of medical and psychological professionals in the U.S. Government regarding interrogation practices and other detention procedures. Physicians for Human Rights has, for example, followed these issues closely and criticized the ethical behavior of these professionals.[19] The Coalition for an Ethical Psychology, a group of psychologists within the

American Psychological Association announced in 2008 that "...the American Psychological Association (APA) passed a referendum banning participation of APA member psychologists in US detention facilities, such as Guantanamo or the CIA's 'black sites' operating outside of or in violation of international law or the Constitution."[20] Another NGO specifically focused on promoting arms control for military robots has recently been formed, called the International Committee for Robot Arms Control (ICRAC).[21]

At the national level in the United States, existing governance can be described as *decentralized*, and in one sense, *reactionary*. It reflects the *push* and *pull* of multiple constituencies and philosophies regarding the efficacy of support for technological innovation. U.S. federal law and regulation reflect the belief that innovation is best encouraged on the one hand by vigorous and unrestrained marketplace competition, while recognizing, on the other hand, the need for the government to organize federal funding, encourage innovation, and regulate the more egregious results of commercialization.[22] The President's Council on Bioethics recognized this fact in 2003 in its *Report on the State of Biotechnology*:

> Whether one likes it or not, progress in biology and biotechnology is now intimately bound up with industry and commerce....Whatever one finally thinks about the relative virtues and vices of contemporary capitalism, it is a fact that progress in science and technology owes much to free enterprise. The possibility of gain adds the fuel of interest to the fire of genius, and even as the profits accrue only to some, the benefits are, at least in principle, available to all. And the competition to succeed provides enormous incentives to innovations, growth, and progress. We have every reason to expect exponential increases in biotechnologies, and, therefore, in their potential uses in all aspect of human life.[23]

Within the U.S., for example, there appears to be no urgency regarding the coordination of governance of emerging technologies within the federal government generally; nor is there any evidence of a prevailing belief that the present governance architecture requires any type of thorough overhaul to respond to the challenges of the 21st century. Indeed the President's Council of Advisors on Science and Technology reported in 2008:

> [T]here are no ethical concerns that are unique to nanotechnology today. That is not to say that nanotechnology does not warrant careful *ethical* evaluation. As with all new science and technology development, all stakeholders have a shared responsibility to carefully evaluate the ethical, legal, and societal

implications raised by novel science and technology developments. However, the[re is]...no apparent need at this time to reinvent fundamental ethical principles or fields, or to develop novel approaches to assessing societal impacts with respect to nanotechnology.[24]

A more recent study, commissioned by DARPA itself to study the ethical, legal, and societal issues regarding emerging and readily available technologies and national security is not so sanguine. It recognized that the policy of technological superiority must still conform to the laws of war and yet technological innovation often runs way a head of the intent if the not the specific strictures of IHL. The report concluded that IHL provides less and less restraint.[25]

Part two: International Humanitarian Law (IHL)

Turning to military uses of technologies for the projection of force, specifically, development and use continue to be constrained, as mentioned above, by various restrictions regarding the projection of force found in international law, as translated variously into national laws and regulations. There are, as cited above, multiple conventions which purport to deal with specific technologies and practices. Even though the United States is not a party to all of these conventions, nor necessarily bound by all of them, it is nonetheless the case that the US has taken considerable interest in the articulation of standards which purport to regulate conduct generally on the battlefield, including how weapons are used.

There are five principles which run through the language of the various humanitarian law treaties (the rules) which the United States acknowledges and generally honors regarding the conduct of warfare. These are: (i) a general prohibition on the employment of weapons of a nature to cause superfluous injury or unnecessary suffering, (ii) military necessity, (iii) proportionality, (iv) discrimination, and (v) command responsibility. These principles, as discussed below, impose ethical and arguably legal restraints on at least some uses of emerging military technologies.

First, some weapons, it is argued, are patently inhumane, no matter how they are used or what the intent of the user is. This principle has been recognized since at least 1907,[26] although consensus over what weapons fall within this category tends to change over time. The concept here is that some weapons are *design-dependent*: that is, their effects are reasonably foreseeable even as they leave the laboratory. In 1996, the

International Committee of the Red Cross at Montreux articulated a test to determine if a particular weapon would be the type which would foreseeably cause superfluous injury or unnecessary suffering.[27] The so-called SIrUS criteria would ban weapons when their use would result in:

- A specific disease, specific abnormal physiological state, a specific and permanent disability or specific disfigurement; or
- Field mortality of more than 25 percent or a hospital mortality of more than 5 percent; or
- Grade 3 wounds as measure by the Red Cross wound classification scale; or
- Effects for which there is no well-recognized and proven treatment.[28]

The operative term here is *specific*; the criteria speak to technology specifically designed to accomplish more than merely render an adversary *hors de combat*. This test for determining weapons exclusion is a medical test and does not take into consideration the issue of military necessity. For this reason, these SIrUS criteria have been roundly criticized and rejected by the United States specifically, and by the international community generally, notwithstanding support for the general principle against the use of inhumane weapons.[29]

The second principle, *military necessity,* requires a different analysis. This principle "…justifies measures of regulated force not forbidden by international law which are indispensable for securing the prompt submission of the enemy, with the least possible expenditures of economic and human resources."[30] It is justified, according to this principle, to project force in order to secure legitimate military objectives which are generally limited to those objects which by their nature, location, purpose or use make an effective contribution to military action and whose total or partial destruction, capture or neutralization, in the circumstances ruling at the time, offers a definite military advantage. *Military necessity* recognizes the benefit to friend and foe alike of a speedy end to hostilities. Protracted warfare, it assumes, creates more rather than less suffering for all sides. In order to determine the necessity for the use of a particular technology, then, one needs to know what the definition of victory is, and how to measure the submission of the enemy in order to determine whether the technology will be *necessary* in this regard.

The third principle, *proportionality,* is of considerable concern to the developer and user of new technologies. A use of a particular technology is not *proportional* if the loss of life and damage to property incidental to attack is excessive in relation to the concrete and direct military advantage expected to be gained.[31] In order to make this determination, it can be argued, one must consider the military necessity of a particular use and evaluate the benefits of that use in furtherance of a specific objective against the collateral damage that may be caused.

Discrimination, the fourth principle, goes to the heart of moral judgment. Indiscrminate attacks (uses) are prohibited under the rules. Indiscriminate uses occur whenever such uses are not directed against a specific military objective, or otherwise employ a method or means of combat the effects of which cannot be directed at a specified military target (indiscriminate bombing of cities, for example). Indiscriminate usage also encompasses any method or means of combat, the effects of which cannot be limited as required, or that are otherwise of a nature to strike military and civilian targets without distinction.

A final principle is *command responsibility*, that principle which exposes a multiple of superiors to various forms of liability for failure to act in the face of foreseeable illegal activities. This is a time-honored principle, grounded on the contract between soldiers and their superiors, which requires soldiers to act and superiors to determine when and how to act. It has a long history reflective of the need for control on the battlefield.[32] Presumably, the illegal development, adaption or use (misuse) of a particular technology is the responsibility of the superior in the chain of command who issues such an order. The defenses of *superior orders* and *defense of the nation* are not available.

A 1997 Protocol to the Geneva Convention requires that each State Party "determine whether the employment of any new weapon, means or method of warfare that it studies, develops, acquires or adopts would, in some or all circumstance, be prohibited by international law."[33] The legal framework for this review is the international law applicable to the State, including IHL. In particular this consists of the treaty and customary prohibitions and restrictions on specific weapons, as well as the general IHL rules applicable to all weapons, means and methods of warfare. General proscriptions include the principles described above, such as protecting civilians from the indiscriminate effects of weapons and combatants from unnecessary suffering. The assessment of a weapon in light of the relevant rules will require an examination of all relevant empirical information pertinent to the weapon, such as its technical description and actual performance, and its effects on health and the environment. This is the rationale for the involvement of experts of various disciplines in the review process.[34]

Once again, the United States is not a signatory to this Protocol and thus, technically not bound by its requirements. Nonetheless, to the extent that it sets out reasonable requirements and methodologies for use by states fielding new and emerging technologies, this treaty could well set the standard in international law for what may be considered appropriate conduct.

A final constraint worth noting is the emerging trend in international law to hold those responsible for fielding weapons which allegedly contravene the principles enunciated above through the use of litigation based on the concept of *universal jurisdiction*. The concept of universal jurisdiction is a customary international law norm that permits states to regulate certain conduct to which they have no discernible nexus. Generally, it is recognized as a principle of international law that all states have the right to regulate certain conduct regardless of the location of the offense or the nationalities of the offender or the victims. Piracy, slave trade, war crimes, and genocide are all generally accepted subjects of universal jurisdiction. Belgium, Germany, and Spain have all entertained such prosecutions and a number of U.S. officials including George W. Bush, Colin Powell, Tommie Franks. Henry Kissinger and Donald Rumsfeld have been named in investigations, although their prosecutions have been without success.

The issue of *lawfare* is also of concern. Lawfare is a strategy of using or misusing law as a substitute for traditional military means to achieve military objectives. Each operation conducted by the U.S. military results in new and expanding efforts by groups and countries to use lawfare to respond to military force. American military authorities are still grappling with many of these issues. While litigation to date has revolved primarily around allegations of practices such as genocide, torture, rendition, and illegal interrogation, there is no reason to believe that future prosecutions may be justified where decisions regarding illegal innovation, adaptation, and use of weapons systems are made.[35]

These various principles and requirements of international humanitarian law and ethical rules of military conduct would clearly impose some limitations on the development and use of emerging military technologies. However, given the ambiguous meaning and uncertain legal binding status of these principles, they are unlikely to adequately constrain and shape the development and use of emerging technologies on their own. Additional oversight mechanisms may therefore be warranted.

Notes

1. See generally Stephen E. White, "Brave New World: Nurowarfare and the Limits of International Humanitarian Law," 41 *Cornell Int'l L.J.* 177 (2008); Mark Edward Peterson, "The UAV and the Current and Future Regulatory Construct for Integration into the National Airspace System," 71 *J. Air Law & Commerce* 521 (2006); Geoffrey S. Corn, "Unarmed but How Dangerous? Civilian Augmentees, The Law of Armed Conflict, and the Search for a More Effective Test for Permissible Civilian Battlefield Functions," 2 *J.Nat'l Security L.& Pol'y* 257 (2008); Andrew H. Henderson, "Murky Waters: The Legal Status of Unmanned Undersea Vehicles," 53 NAVAL L. REV. 55 (2006); Jason Borenstein, "The Ethics of Autonomous Military Robots," 2 *Studies in Ethics, Law & Tech.* Issue 1, Article 2 (2008); John J. Klein, "The Problematic Nexus: Where Unmanned Combat Air Vehicles and the Law of Armed Conflict Meet," *Air & Space Power J. Chronicles Online* J (2003), retrieved at http://www.airpower.maxwell.af.mil/airchronicles/cc/klein.html, October 20, 2010. Anthony J. Lazarski, "Legal Implications of the Uninhabited Combat Aerial Vehicle—Focus: Unmanned Aerial Vehicles," 16 *Aerospace Power J.* 74 (2002).
2. Convention on the Prohibition of the Development, Production and Stockpiling of Bacteriological (Biological) and Toxin Weapons and or their Destruction (1972), 26 U.S.T. 583, 1015 U.N.T.S. 163.
3. Convention on the Prohibition of the Development, Production, stockpiling and use of Chemical Weapons and on Their Destruction, January 13, 1993, 1974 U.N.T.S. 45.
4. The 1999 Hague Declaration Concerning Expanding Bullets, July 29, 1899, 1. AM. J. *Int'l L* 157–59 (Supp.).
5. Convention on the Prohibition of Military or Any Hostile Use of Environmental Modification Techniques, May 18, 1977, 31 U.S.T. 333 16 I.L.M. 88.
6. Protocol on Prohibitions or Restrictions on the Use of Mines, Booby-Traps and Other Devices As Amended on May 3, 1996 (Protocol II As Amended on May 3, 1996) Annexed to the Convention on Prohibitions or Restrictions on the Use of Certain Conventional Weapons Which May Be Deemed to Be Excessively Injurious or to Have Indiscriminate Effects, May 3, 1996, 2048 U.N.T.S. 133; Convention on the Prohibition of the Use, Stockpiling, Production and Transfer of Anti-Personnel Mines and on their Destruction, September 18, 1997, 2056 U.N.T.S. 211.

7   Protocol on Prohibitions or Restrictions on the Use of Incendiary Weapons, October 10, 1980, 1342 U.N.T.S. 171.
8   Protocol on Blinding Laser Weapons (Protocol IV), October 13, 1995, 35 I.L.M. 1218.
9   See generally International Committee of the Red Cross, International Humanitarian Law-treaties and Documents, retrieved at http://www.icrc.org/ihl.nsf/TOPICS?OpenView, November 30, 2009.
10 There has been a good deal of discussion in recent years about the subject of good governance, especially in the development area. The United Nations, for example, lists eight characteristics of good governance, which are: consensus oriented, participatory, adherence to the rule of law, effect and efficient, accountable, transparent, responsive, equitable and inclusive. United Nations Economic and Social Commission for Asia and the Pacific, "What Is Good Governance?" United Nations, 2009, retrieved at http://www.unescap.org/pdd/prs/ProjectActivities/Ongoing/gg/governance.asp, October 20, 2010; see also Sam Agere, Good Governance, Promoting Good Governance; Principles, Practices and Perspectives (London: Marlboro House, 2000).
11 APA Ethics Code, 2006. "Introduction to American Psychological Association Ethics Code," retrieved at http://www.apa.org/ethics/code2002.html, November 25, 2009.
12 APA Ethics Code "Standard 1:02 Conflicts Between Ethics and Law, Regulations or Other Governing Legal Authority".
13 American Medical Association, "American Medical Association Principles of Medical Ethics (Principle (1))," retrieved at http://www.ama-assn.org/ama/pub/physician-resources/medical-ethics/code-medical-ethics/principles-medical-ethics.page
14 American Society of Civil Engineers, "American Society of Civil Engineers, Code of Ethics," retrieved at http://www.asce.org/inside/codeofethics.cfm, November 24, 2009.
15 ISACA, ISACA Code of Professional Ethics, retrieved at http://www.isaca.org/Template.cfm?Section=Code_of_Ethics1&Template=/TaggedPage/T, November 25, 2009.
16 Republic of Korea, Ministry of Information and Communication as quoted in Stefan Lovgren "Robot Code of Ethics to Prevent Android Abuse, Protect Humans," National Geographic News, March 16, 2007, retrieved at http://news.nationalgeographic.com/news/2007/03/070316-robot-ethics.html, November 26, 2009.
17 Pledge of Ethical Conduct, University of California, Berkley, 1998, retrieved at http://courses.cs.vt.edu/cs3604/lib/WorldCodes/Pledge.html, January 15, 2010.
18 IPEN. "Welcome to the International POPs Elimination Network," retrieved at http://www.ipen.org/, October 10, 2010.
19 Physicians for Human Rights, White Paper, "Broken Laws, Broken Lives" (November 28, 2007).

20 Stephen Soldz, "Ending the Psychological Mind Games on Detainees," *Op-ed Boston Globe*, August 14, 2008, retrieved at http://brokenlives. info/?tag=psychologifsts, November 30, 2009. See also, Institute on Medicine as a Profession, *Ethics Abandoned, Medical Professionalism and Detainee Abuse in the War on Terror* (New York: Colombia University, 2013) and Hernan Reyes, Scott A. Allen, George J. Annas, "Physicians and Hunger Strikes in Prison: Confrontation, Manipulation, Medicalization and Medical Ethics," *World Medical Journal 59*, no. 1 (February 2013), and no. 2 (April 2013).
21 http://www.icrac.co.cc/, October 04, 2010.
22 See, for example, Harris-Kefauver Act. Pub. L. No. 87–781. 76 Stat. 780 amending 21 U.S.C. *sec.* 301 *et seq.* (1962) [commonly referred to as the 1962 Drug Amendments]; the National Research Act of 1974 (P.L. 93–348); and the 21st Century Nanotechnology Research and Development Act (P.L. 108–153, 15 USC 7501 *et seq.*).
23 The President's Council on Bioethics. "Beyond Therapy: Biotechnology and the Pursuit of Happiness," 303.
24 President's Council of Advisors on Science and Technology, National Nanotechnology Initiative, "Second Assessment and Recommendations of the NNAP" (April 2008), retrieved at www.ostp.gov , January 15, 2010.
25 Chameau, Ballhaus and Lin, *Emerging and Readily Available Technologies and National Security*, 40.
26 *See* International Conferences (The Hague), Hague Convention (IV) Respecting the Laws and Customs of War on Land and Its Annex: Regulations Concerning the Laws and Customs of War on Land, October 18, 1907, retrieved at http://www.unhcr.org/refworld/docid/4374cae64.html , October 9, 2010.
27 International Committee of the Red Cross, "The Medical Profession and the Effects of Weapons, Symposium" (Montreux, Switzerland, 1996).
28 International Committee of the Red Cross, "The SIrUS Project: Towards a Determination of Which Weapons Cause 'Superfluous Injury or Unnecessary Suffering,'" ICRC, Geneva, 1997. See also Andrew Kock, "Should War Be Hell?" *Jane's Defense Weekly* (May 10, 2000), 23.
29 Donna Marie Verchio, "Just Say No! The SIrUS Project: Well-intentioned, but Unnecessary and Superfluous," 51 *The Air Force Law Review* 183 (2001).
30 Roy Gutman & Daoud Kuttab, *Indiscriminate Attack*, in *Crimes of War: What the Public Should Know* (2007), retrieved at http://www.crimesofwar.org/thebook/indiscriminate-attack.html, October 9, 2010.
31 U.S. Army Field Manual 27-10, The Law of Land Warfare, para. 41, change 1 (1976), retrieved at http://www.globalsecurity.org/military/library/policy/army/fm/27-1-/, November 10, 2009.
32 Brandy Womack, "The Development and Recent Applications of the Doctrine of Command Responsibility: With Particular Reference to the Mens

Rea Requirement," in Yee Sienho (ed.), *International Crime and Punishment*, Selected issues 117 (2003).

33 Protocol Additional to the Geneva Conventions of August 12, 1949, and relating to the Protection of Victims of International Armed Conflicts, June 8, 1977, Article 36 of 1977.

34 Kathleen Lewand, "A Guide to the Legal Review of New Weapons, Means and Methods of Warfare, Measure to Implement Article 36 of Additional Protocol 1 of 1977," International Committee of the Red Cross Publication 0902 (2007), retrieved at http://www.icrc.org/web/eng/siteengo.nsf/html/p0902, October 10, 2010.

35 Council on Foreign Affairs, Transcript, "Lawfare, The Latest in Asymetrics," March 18, 2003, retrieved at http://www.cfr.publications.html?id=5772, December 10, 2009.

6
Arms around the Problem: Suggestions for Future Governance

Abstract: *The decision when to design, what to design, and how to use a particular technology is routinely left up to the engineers who work with the science upon which the technologies are based. There is a wide array of additional specialties including non-governmental organizations, international organizations, non-state actors, and those who are impacted by their use whose concerns need to be considered. For the military, decisions regarding design and use of these technologies is of particular and immediate concern. Governance of military technologies must reflect the legal and ethical concerns of the people the military is sworn to protect; yet it must also recognize the existential need for soldiers to accomplish a myriad of violent and dangerous tasks while at the same time looking out for the welfare of the soldier.*

> *Part one: Who gets to decide?*
> *Part two: How to regulate?*

O'Meara, Richard Michael. *Governing Military Technologies in the 21st Century.* New York: Palgrave Macmillan, 2014. DOI: 10.1057/9781137449177.0010.

As has been described above, emerging technologies are innovated, adapted, and used in a relatively freewheeling commercial environment, encouraged by the exigencies of globalized markets, fueled by multiple agenda and justified by the assumption that more technology means more progress. Each technology, especially biotechnology, cyber technology, and robotics carries with it its own set of consequences, good and bad, and all are beset with the probability of considerable unanticipated consequences for the future.

All technology, in one sense, can be viewed as enhancement in that it enables humans to achieve certain effects that would otherwise require more effort or be altogether impossible to obtain without it. As Nick Bostrom and Julian Savulescu observe "[M]any of the ethical issues that arise in the examination of human enhancement prospects hook into concepts...such as human nature, personal identity, moral status, well-being, and problems in normative ethics, political philosophy, philosophy of mind, and epistemology. In addition to these philosophical linkages, human enhancement also offers thought-fodder for several other disciplines, including medicine, law, psychology, economics, and sociology."[1]

The need for governance, if it exists, is bottomed on the uses to which technology can be put and the consequences which may occur if no governance is forthcoming. Moreno, for example, concludes that the "...proper response to transhumanism is not to prohibit research and development of these new technologies but to develop careful monitoring and regulatory systems."[2] Singer and Krishnan, amongst many others, agree.[3] Despite Kurzweil's optimism and trust in the nature of technological innovation,[4] it can be argued that the cost of not attempting regulation of some kind may be too great to bear.[5] The first question, then, is who should do the regulating?

Part one: Who gets to decide?

According to Fukuyama, the debate which has occurred regarding regulation of these new technologies has been considerably polarized to little effect. The state of the debate on biotechnology, for example "...is today polarized between two camps."

The first is libertarian and argues that society should not and cannot put constraints on the development of new technology. This camp includes researchers and scientists who want to push back the frontiers of science, the biotech industry that stands to profit from unfettered technological advance, and, particularly in the United States and Britain, a large group that is ideologically committed to some combination of free markets, deregulation, and minimal government interference.

The other camp is a heterogeneous group with moral concerns about biotechnology, consisting of those who have religious convictions, environmentalists, with a belief in the sanctity of nature, opponents of new technology, and people on the Left who are worried about the possibility of eugenics.[6]

Oversight, including the ethical discussions regarding the efficacy of innovation has, again according to Fukuyama, yielded minimal results. He notes, for example, that "[I]n any discussion of cloning, stem cell research, germ-line engineering, and the like, it is usually the professional bioethicist who can be relied on to take the most permissive position of anyone in the room." He ascribes this phenomenon to a cultural condition he refers to as *regulatory capture*, whereby

> the group that is supposed to be overseeing the activities of an industry becomes an agent for the industry. This happens for many reasons, including the dependence of the regulators on the regulatees for money and information. In addition, there are the career incentives that most professional bioethicists face. Scientists do not usually have to worry about winning the respect of ethicists, particularly if they are Nobel Prize winners in molecular biology or physiology. On the other hand, ethicists face an uphill struggle winning the respect of the scientists they must deal with, and are hardly likely to do so if they tell them they are morally wrong or if they depart significantly from the materialist worldview that the scientists hold dear.[7]

Moreno's concern is a different one. He notes that a good deal of the most dangerous technological innovation occurs not in the transparent world of Francis Bacon's "scientific community" but rather in the classified, highly secretive environment of the national security state.[8]

No matter the reason for the danger, they both suggest that it is the policy-makers; the representatives of those who will be most affected by the consequences of technological innovative, who bear responsibility for regulation.[9]

> It is only "theology, philosophy, or politics" that can establish the ends of science and the technology that science produces, and pronounce on whether those ends are good or bad. Scientists may help establish moral rules concerning their own conduct, but they do so not as scientists but as scientifically informed

members of a broader political community. There are very many brilliant, dedicated, energetic, ethical, and thoughtful people within the community of research scientists....But their interests do not necessarily correspond to the public interest. Scientists are strongly driven by ambition, and often have pecuniary interests in a particular technology or medicine as well.

Hence the question of what we do with biotechnology is a political issue that cannot be decided technocratically.[10]

This same issue occurs in the military, which after all is in part a bureaucratized organization subject to many of the same pushes and pulls that inform other organizations driven by technological innovation. Routinely, lawyers are used to accomplish this task of advice when discussions occur regarding the ethical efficacy of a particular practice or technology. Lawyers in the military have been referred to as *the conscious of the command*[11] and yet, they are neither trained nor philosophically situated to accomplish this task. Their careers are in the hands of those whom they are called upon to regulate and they are often susceptible to being captured in the same manner as Fukuyama's bioethicists.

When Tommie Franks, during the first days of the war in Iraq and Afghanistan, made targeting decisions [choices regarding the use of particular technologies to obtain particular results] based on his lawyer, "My JAG [Judge Advocate General] doesn't like this, so we're not going to fire"[12] he was arguably abrogating his responsibility to decide when to project force. His lawyer could tell him what he could *legally* do but not what he *should* do. That, of course, is an ethical decision. It requires a separate set of skills, a whole host of mature experiences, and a very different way of looking at the world to make that decision correctly. Policy makers—those involved in determining *if* a technology should be developed, *when* it should be developed, *how* it should be developed, *who* should use it and for *what purpose*—consider legal determinations but must go well beyond them. The law is a conservative animal, often reactive and well behind the contemporary problem. It, perforce, must look backward to precedent and place the present conundrum within the confines of what has gone before. It is, ultimately, bottomed on rationality, what is reasonable under the circumstance. On occasion, however, reason can miss the mark. In addition, it does not consider answers to a whole host of other questions which come up in the environment of the new battlefield.[13] One is reminded of the *Star Trek* episode summarized by Gutlein as follows: Captain Kirk encounters a world where war is waged by computers and probabilities. The worlds of Eminar VII and Vendikar have been at war for over 500 years. The two planets have learned to avoid the horrors of war by the use of computers. When the computers score a "hit," casualty estimations are made, and people are ordered to disintegration chambers to be atomized. Captain Kirk is appalled by the scientific [rational?] approach to warfare. *They have made this war too easy and until they experience the horrors of war, there will never be any incentive to make peace* [emphasis added].[14]

If war is too important to be left to soldiers, as Clemenceau is famously quoted as observing, technology may well be too important to be left to scientists and lawyers.

Part two: How to regulate?

When confronted with the considerable potentialities posed by emerging military technologies, their rapid innovation, adaption and use on the battlefield, one is tempted to throw up one's hands, declare all the old rules dead, and begin anew to draft proscriptions that comport to some new 21st-century logic. It can be argued that this would be a mistake.

There are a host of models in international law which may be useful should one wish to seek international regulation of the various specific issues which each technology brings to the table.

a. International treaties

A more formal and traditional approach for oversight of a new weapons system would be some form of binding international agreement. Under existing international law, there are a significant number and diversity of precedents for restricting specific weapons. As has been demonstrated, existing legally binding arms control agreements and other instruments include a wide variety of different types of restrictions on targeted weapons, including prohibitions and limitations (restrictions that fall short of prohibition) on acquisition, research and development, testing, deployment, transfer or proliferation, and use.

These various types of prohibitions and limitations form a kind of menu from which the drafters of an international legal instrument addressing emerging military technologies, designed to project force either mechanistically or through the enhancement of individual soldiers, could choose in accordance with their goals and the parameters of political support for such restrictions. A similar menu could be created of the various types of monitoring, verification, dispute resolution, and enforcement mechanisms that implement the prohibitions and limitations contained in existing international legal arms control instruments.

These prohibitions and limitations (as well as any accompanying monitoring/verification, dispute resolution, and enforcement provisions) can be contained in any of a number of different types of international legal instruments. They are typically contained in legally binding multilateral agreements, included in multilateral agreements primarily focused on arms control and also in the Rome Statute of the International Criminal Court. However, there are also examples of prohibitions and limitations contained in legally binding bilateral agreements as well as examples of prohibitions and limitations contained in legally binding resolutions of the United Nations Security Council or in customary international law (which consists of rules of law derived from the consistent conduct of States acting out of the belief that the law required them to act that way).

New international legal arms control instruments are typically free-standing. However, there is also at least one existing multilateral legal framework agreement which might be amended to itself provide a vehicle for some or all desired restrictions on military technologies, especially lethal autonomous robots (LARs). This is the 1980 Convention on Prohibitions or Restrictions on the Use of Certain Conventional Weapons which may be deemed to be Excessively Injurious or to have Indiscriminate Effects (the CCW),[15] which has been ratified by over 100 states parties.[16]

The operative provisions of the CCW are contained within its protocols. The five protocols currently in force contain rules for the protection of military personnel and, particularly civilians and civilian objects from injury or attack under various conditions by means of fragments that cannot readily be detected in the human body by x-rays (Protocol I), landmines and booby traps (amended Protocol II), incendiary weapons (Protocol III), blinding lasers (Protocol IV), and explosive remnants of war (Protocol V).[17] Prohibition against the use of LARs, for example, might well fall within the proscription of the CCW preamble "that prohibits the employment in armed conflicts of weapons, projectiles and material and methods of warfare of a nature to cause superfluous injury or unnecessary suffering."[18]

Some international legal arms control agreements prohibit a full range of activities involving targeted weapons. For example, states—parties to the Convention on the Prohibition of the Use, Stockpiling, Production and Transfer of Anti-Personnel Mines and on Their Destruction—typically referred to as the "Mine Ban Treaty,"—commit to not developing, producing, acquiring, retaining, stockpiling, or transferring anti-personnel landmines.[19] The following menu contains additional examples of existing international legal instruments which adopt specified types of restrictions on a narrower basis.

1. Prohibitions and limitations on the acquisition of certain weapons

Several international legal arms control instruments completely prohibit the acquisition of targeted weapons. For example, the Biological Weapons Convention (BWC) prohibits all state-parties from acquiring, producing, developing, stockpiling, or retaining—and requires all state-parties to, within nine months, destroy or divert to peaceful purposes—i) biological agents and toxins "...of types and in quantities that have no justification for prophylactic, protective or other peaceful purposes;" and ii) weapons, equipment and delivery vehicles "designed to use such agents or toxins for hostile weapons or in armed conflict."[20] The Convention on the Prohibition of the Development, Production, Stockpiling and Use of Chemical Weapons and on their Destruction (CWC) prohibits all state parties from producing or acquiring, as well as developing, stockpiling or retaining, chemical weapons.[21]

In contrast, the Treaty on the Non-Proliferation of Nuclear Weapons (NPT) creates two classes of states with regard to nuclear weapons.[22] Nuclear-weapon state-parties are those which had manufactured and exploded a nuclear weapon or other nuclear explosive device prior to January 1, 1967 (China, France, Russia, the United Kingdom, and the United States).[23] The NPT does not require nuclear-weapon state-parties to give up their nuclear weapons, but does require those parties to "...pursue negotiations in good faith on effective measures relating to cessation of the nuclear arms race at an early date and to nuclear disarmament."[24] Non-nuclear-weapon state-parties to the NPT are prohibited from receiving, manufacturing, or otherwise acquiring nuclear weapons.[25]

The Inter-American Convention on Transparency in Conventional Weapons Acquisitions[26] provides a very different model, with a focus on transparency rather than prohibition of acquisitions. The Convention does not prohibit any acquisitions but does require its states-parties to annually report on their imports of certain specified heavy weapons, as well as to submit notifications within 90 days of their incorporation of certain specified heavy weapons into their armed forces inventory, whether those weapons were imported or produced domestically.[27]

2. Prohibitions and limitations on research and development

To date, there has been minimal agreement regarding limitations on research and development. One treaty, the CWC, does prohibit the development of all chemical weapon munitions and devices.[28] In contrast, the BWC contains a more nuanced prohibition, banning the development, production, acquisition, and retention of i) microbial or other biological agents or toxins "... of types and in quantities that have no justification for prophylactic, protective or other peaceful purposes" and ii) weapons, equipment or means of delivery "... designed to use such agents or toxins for hostile purposes or in armed conflict."[29] It is important to note that restrictions based on quantities or intended use rather than the underlying nature of the technology can be exceptionally difficult to verify, at least without highly intrusive inspections.

3. Prohibitions and limitations on testing

Prohibitions and limitations on testing of targeted weapons are most prominent in the nuclear weapons context. For example, the Comprehensive Test Ban Treaty (CTBT), which has not yet entered into force, prohibits "... any nuclear weapon test explosion or any other nuclear explosion."[30] In contrast, the 1963 Treaty Banning Nuclear Weapon Tests in the Atmosphere, in Outer Space and Under Water (also known as the "Limited Test Ban Treaty")—which unlike the CTBT is in force— specifically prohibits nuclear weapons tests "... or any other nuclear explosion" not only in the atmosphere but in outer space, and under water.[31] The Limited Test Ban Treaty also prohibits nuclear explosions in all other environments, including underground, if they cause "... radioactive debris to be present outside the territorial limits of the State under whose jurisdiction or control" the explosions were conducted.[32]

4. Prohibitions and limitations on deployment

Some international legal arms control instruments focus on limiting deployment of targeted weapons, using targeted caps as the limiting factor. For example, the Strategic Offensive Reductions Treaty, entered into by the U.S. and Russia in 2002, requires the two countries to reduce their operationally deployed strategic nuclear forces to between 1,700 and 2,200 warheads by December 31, 2012.[33] The Conventional Armed Forces in Europe Treaty, ratified by the United States in 1992, contains bloc and regional limits on deployment of certain weapons as well.[34]

5. Prohibitions and limitations on transfer/proliferation

Many international legal arms control instruments include prohibitions or limitations on transfer or other proliferation of the targeted weapons. For example, the NPT prohibits parties that possess nuclear weapons from transferring the weapons to any recipient as well as from assisting, encouraging, or inducing any non-nuclear-weapon state to manufacture or otherwise acquire such weapons in any way.[35]

The CWC bans the direct or indirect transfer of chemical weapons.[36] The CWC also bans assisting, encouraging, or inducing anyone to engage in CWC-prohibited activity.[37] Similarly, the BWC bans the transfer to any recipient, directly or indirectly, and assisting any state, group of states, or international organizations to manufacture or otherwise acquire i) biological agents and toxins "…of types and in quantities that have no justification for prophylactic, protective or other peaceful purposes;" and ii) weapons, equipment and delivery vehicles "…designed to use such agents or toxins for hostile weapons or in armed conflict."[38] In contrast, the Inter-American Convention on Transparency in Conventional Weapons Acquisitions does not prohibit exports but does require its states-parties to annually report on their exports of certain specified heavy weapons.[39]

6. Prohibitions and limitations on use

Several international treaties include prohibitions or limitations on use of targeted weapons. The International Court of Justice, in a 1996 advisory opinion on the Legality of the Threat or Use of Nuclear Weapons, ruled that "…the threat or use of nuclear weapons would generally be contrary to the rules of international law applicable in armed conflict, and in particular the principles and rules of humanitarian law; however, in view of the current state of international law, and of the elements of fact at its disposal, the Court cannot conclude definitively whether the threat or use of nuclear weapons would be lawful or unlawful in an extreme circumstance of self-defense, in which the very survival of a State would be at stake."[40] The Rome Statute of the International Criminal Court prohibits i) employing poison or poisoned weapons, ii) employing poisonous gases, and iii) employing bullets which flatten or expand easily in the human body.[41] This list is potentially expandable. While the CWC bans chemical weapons use or military preparation for use,[42] the BWC does not ban the use of biological and toxin weapons but reaffirms the 1925 Geneva Protocol, which prohibits such use.[43]

Protocol IV of the CCW prohibits the use of lasers specifically designed to cause permanent blindness.[44] It further obliges states-parties to make every effort to avoid causing permanent blindness through the use of other lasers.[45] While prohibiting the use of blinding lasers, the convention does not rule out their development or stockpiling.[46] However, it does outlaw any trade in such arms.

As models for additional agreements regarding emerging military technologies, international treaties present both strengths and weaknesses. Discussions of the efficacy of international law often divide the analysis into articulation, institutionalization, and enforcement. Clearly, it can be argued, it is in the area of articulation that international law (especially treaty law) provides its greatest service. Treaties do not spring from whole cloth; rather they come about over the course of time and often represent attempts by stakeholders, especially but not always states, to reach consensus on important issues. Jack Goldsmith and Eric Posner find that international agreements provide methodologies for cooperation, coordination, and identification of coincidence of interest which, perhaps, did not exist before, and, further, provide considerable information for stakeholders which aids them in pursuing courses of action going forward.[47] Some commentators believe that treaties create habits of governance, a sort of political will, which creates obligations for compliance.[48] Goldsmith and Posner are less sanguine about the ability of international treaty regimes to enforce proscriptions and conclude "[I]nternational law is a real phenomenon, but international law scholars exaggerate its power and significance. We have argued that the best explanation for when and why states comply with international law is not that states have internalized international law, or have a habit of complying with it, or are drawn to its moral pull, but simply that states act out of self-interest."[49]

Further, it should be emphasized that international treaties run bilaterally or multilaterally between states and, therefore, do not purport to regulate or constrain the multiple additional actors on the international stage, such as terrorist organizations, criminal syndicates, non-governmental organizations, private military contractors and international corporations. Given that emerging military technologies, especially in the cyber and biotechnologies are routinely dual-use and innovated in private spheres, enforcement by international institutions is problematic. Finally, there is a good deal of evidence to support the proposition that states routinely ignore without consequence those portions of international treaties which are inconvenient or which they perceive to be inconsistent with national interests. Gross goes so far as to take the position that acquiesce in conduct by international actors which is violative of traditional IHL and other portions of the Human Rights regime constitutes norm changes within the definition of customary international law. He concludes "[B]ased on state behavior and the substance of the arguments that justify rather than excuse, there is preliminary evidence that targeted killings, aggressive interrogation, non-lethal weapons, and attacks on participating civilians (by either side) reflect emerging norms of warfare. Whether these norms are new rules or acceptable exceptions, they are far from the prohibitions and severe restrictions that currently characterize the laws of war."[50]

International treaties, then, have the ability to sharpen the discussion and identify agreed upon standards of conduct. They often set up institutions to monitor the conduct of the obligors and, where there is a strong utilitarian interest, obtain enforcement. They work best when there is a strong utilitarian interest to seek compliance and when all stakeholders are involved and fail when there is poor articulation or where the stated goals are vague and unenforceable.

b. The case for IHL (The Law of War)

As noted above, IHL is a set of treaties and international obligations which purport to regulate the conduct of force projection during the conduct of war. It should be remembered that IHL is the product of centuries of experience. The *genie* has been out of the bottle before and demonstrated to mankind in the first half of the 20th century the ramifications of minimal rules and inattention to governance. For all their failures, the Geneva Conventions of 1949, the Human Rights Regime, and the subsequent conventions which purport to address specific issues regarding the projection of force (the rules) have been created at special moments in history, when states were prepared, for whatever reason, to acknowledge the failure of unilateral power to order or at least constrain the horrors of the battlefield. For the realist, IHL is only the recognition that some sense of governance has its usefulness; for the idealist there is a hope that man can learn something from death and dying on a massive scale. No matter the disagreement, the rules have had their benefits.[51] Unlike many international treaty proscriptions, IHL is firmly grounded in utilitarian concerns, for example protection where possible of all who find themselves on the battlefield.

Clearly the environment in which warfare is conducted has changed radically. On the other hand, there is still the possibility of armed conflicted between nation states doing battle in defined geographies with relatively symmetrical weapons systems. The first Gulf War is an example of this type of conflict; nation-states fighting with uniformed soldiers, constrained at least in part from using all the weapons available—no gas, no anthrax, no revenge or retribution. Thousands of Iraqi soldiers and a number of American soldiers benefited from rules regarding the treatment of prisoners of war in this conflict. Even the United Nations ultimately gave this war its blessing. There is no reason to believe that in the 21st century, this type of conflict will not occur again. In a number of other conflicts, Vietnam, Kosovo, and the Second Iraq War, at least one side—the United States—has seen fit to conduct itself in relative compliance with the rules. Again, for the soldiers on the ground, especially enemy soldiers, or the pilots taken prisoner, these rules have often had their benefits.

Yet, as Gross and others counsel, there has been an increasing drift toward warfare in which at least one party denies the relevance of the rules completely, fights asymmetrically using all manner of weapons and practices that are clearly prohibited. Civilians are the biggest targets, the ultimate losers in these conflicts. These conflicts are fought amongst them; they are targeted and terrorized. The ability to get at them becomes proof that adversaries are powerful, capable of delegitimizing the security efforts of governments. These conflicts are fought in a fish bowl; media coverage and the NGO industry are big business, and the soldier's conduct provides the justification for their work. Further, the American soldier in Iraq, Afghanistan, Colombia, Yemen, the Philippines, Somalia, and elsewhere[52] can have no expectation that he will be treated in compliance with the rules. Rather, he can expect to be beaten and beheaded on camera if captured; he can expect all manner of perfidy, use of civilians as shields, suicide attacks, retribution, mercenaries, spies, and disrespect for medical personnel. Conduct on the battlefield is less constrained than at any time since World War II; the adversaries are no longer nation states and are committed to using whatever tools are available to outlast the American soldier until the American public tires of the conflict, and he goes home. Then, if 9/11 provides any example, the enemy will follow the soldier to his house or favorite shopping mall and kill him and his loved ones there. There is nihilism about all this that denies the rationality of the rules and leads the innovator, adaptor, and user of technologies to ask, why follow the rules anyway?

Soldiers strap on not just new technologies when they confront these enemies but new responsibilities in the manner in which they operate. They are required to embrace the *warrior-builder-diplomat spirit*[53] which incorporates the humanitarian justification for their actions. General David Petraeus defined this strategy in a letter to soldiers in Iraq when he assumed command there in 2007:

> I also want you to be aware of my recognition that our focus on securing the population means that many of you will live in the neighborhoods you're securing. That is, in fact, the right way to secure the population—and it means that you will, in some cases, operate in more austere conditions than you did before we adjusted our mission and focus. Rest assured that we will do everything we can to support you as we implement the new plans. This approach is necessary, because we can't commute to the fight in counterinsurgency operations; rather, we have to live with the population we are securing. As you carry out the new approach, I also count on each of you to embrace the warrior-builder-diplomat spirit as we grapple with the demands that securing the population and helping it rebuild will require.

Technology which is permitted to operate outside that spirit is at loggerheads with the new strategy, alienates precisely the people the soldier is sent to secure, and ultimately defeats the purpose of the projection of force. The definition of victory here is measured by adherence to humanitarian principles rather than in spite of them. This is the logic of counterinsurgency operations.

> FMI 3-07-22 Counterinsurgency Operations; Section VI-Rules of Engagement; 2–66.
>
> The proper application of force is a critical component to any successful counterinsurgency operation. In a counterinsurgency, the center of gravity is public support. In order to defeat an insurgent force, US forces must be able to separate insurgents from the population. At the same time, US forces must conduct themselves in a manner that enables them to maintain popular domestic support. Excessive or indiscriminant use of force is likely to alienate the locate populace, thereby increasing support for insurgent forces. Insufficient use of force results in increased risks to US and multinational forces

and perceived weaknesses that can jeopardize the mission by emboldening insurgents and undermining domestic popular support. Achieving the appropriate balance requires a thorough understanding of the nature and causes of the insurgency, the end state, and the military's rule in a counterinsurgency operation. Nevertheless, US forces always retain the right to use necessary and proportional force for individual and unit self-defense in response to a hostile act or demonstrated hostile intent.[54]

The rules take no position regarding the justice of any particular conflict, *jus ad bellum*, but rather speak to how soldiers conduct themselves while involved, *jus in bello*. They assume that wars will end and that the level of enmity that exists during the peace, indeed, the potential for the peace to last, will be based, in part, on the manner in which the parties conducted themselves during the war. Lingering hatred between the adversaries based on the manner in which they fought can corrode a peace and form the basis for new conflict. A second reason for the rules involves the psychological morale of the soldiers themselves. The ideological underpinnings of soldiers who fight these wars on the side of democratic states matter. John McCain has famously made this point:

> This is the destiny of democracy, as not all means are acceptable to it, and all practices employed by its enemies are open before it. Although a democracy must often fight with one hand tied behind its back, it nonetheless has the upper hand. Preserving the Rule of Law and recognition of an individual's liberty constitutes an important component in its understanding of security. At the end of the day, they strengthen its spirit and allow it to overcome difficulties...
> The enemy we fight has no respect for human life or human rights. They don't deserve our sympathy. But this isn't about who they are. This is about who we are. These are the values that distinguish us from our enemies, and we can never, never allow our enemies to take those values away.[55]

As has been discussed above, IHL has the strong utilitarian purpose of protection of multiple participants, voluntary and involuntary, on the battlefield, no matter how that term is defined. Public diplomacy, maintenance of good order and discipline inside and outside the battlespace, and articulation of cultural ethical considerations provide only a short list of its virtues. The argument can be made that rather than discarding the rules in the face of new challenges presented by emerging military technologies, consideration should be given to revisiting them with an eye toward amendment and redefinition.

c. Regulation by the nation-state

Fukuyama makes the case that regulation of emerging technologies including military technologies is, ultimately, a political exercise. "What is important to recognize is that this challenge is not merely an ethical one but a political one as well. For it will be the political decisions that we make in the next few years concerning our relationship to this technology that determines whether or not we enter into a post-human future and the potential moral chasm that such a future opens before us."[56]

Regarding the military, specifically, there is the ever-present tension between multiple sets of concerns, including seeing to the safety of soldiers, accomplishing myriad and diverse missions on behalf of the state with the projection of force, and remaining competitive on the battlefield against adversaries who may operate with a different calculus regarding use of weapons systems and so on. These are clearly political decisions made by civilian authorities within the context of the nation-state. They will, however, be decided as part of a larger discussion regarding the regulation of technology generally.

The state, for all its frailties, is particularly set up to have this conversation and enforce its decisions. First, it has the infrastructure to do so. This infrastructure resides in the political institutions, bureaucracies, and regulatory bodies, private and public, which presently monitor and regulate multiple aspects of society. In this infrastructure, all stakeholders are able to meet and work out the various interests which much be represented for enforceable decisions to be reached. Second, the state is best able, once it has come to a position, to speak to other groups on the international stage and reach consensus, first on a regional basis and then globally. The European Union, for example, has had some success with regard to regulation of technologies, especially in the areas of agriculture. And, clearly it has been the state, specifically the United States and other nuclear powers, which has insured that there has been no unanticipated use of nuclear technology in the last half century. Finally, the state has the ability to enforce its determinations through a whole host of mechanisms including law enforcement, administrative regulation, allocation of resources and leadership.

The international governance of human biotechnology does not inevitably mean the creation of a new international organization, expanding the United Nations, or setting up an unaccountable bureaucracy. At the simplest level it can come about through the effort of nation-states to harmonize their regulatory policies.[57]

Ferguson and Mansbach conclude that "'internationalism' is an orientation to governance and policymaking that is still rooted in a familiar and limited conception of interstate relations."[58] States, then, may still represent the best organizational model to accomplish the politics of regulation regarding emerging military technologies.

Notes

1. Nick Bostrom, Julian Savulescu, "Introduction, Human Enhancement Ethics: The State of the Debate" in Julian Savulescu, Nick Bostrom, eds. *Human Enhancement* (Oxford: Oxford University Press, 2009), 2.
2. Moreno, *Mind Wars*, 135–36.
3. Singer, *Wired for War*, 435; Krishnan, *Killer Robots*, 156. "Most importantly, regulation could prevent an environment that could result in the development of self-evolving powerful autonomous defense systems that could threaten (in the long term) the continued existence of humanity."
4. Kurzweil, *The Singularity Is Near*, 430–31.
5. Joy, "Why the World Doesn't Need Us".
6. Fukuyama, *Our Posthuman World*, 182–183.
7. Ibid., 204.
8. Moreno, *Mind Wars*, 30.
9. Ibid., 185.
10. Ibid., 186.
11. William G. Exhart, "Lawyering for Uncle Sam When He Draws His Sword," retrieved at www.law.umkc.cau/faculty/projects/trials...exhart.html, November 23, 2009.
12. General Tommie Franks quoted in John J. Klein, "The Problematic Nexus: Where Unmanned Combat Air Vehicles and the Law of Armed Conflict Meet," *Air & Space Power Journal-Chronicles on Line Journal*, July 22, 2003, p. 2, retrieved at http://www.airpower.maxwell.af.mil/airchronicles/cc/klein.html, November 08, 2009.
13. For a particular example of what can go wrong when the legal profession invades and dominates the world of policy makers see Jack Goldsmith, *The Terror Presidency, Law and Judgment Inside the Bush Administration* (New York: W.W. Norton, 2007).
14. Guetlein, "Lethal Autonomous Weapons," 14–15.
15. Convention on Prohibitions or Restrictions on the Use of Certain Conventional Weapons Which May Be Deemed to Be Excessively Injurious or to Have Indiscriminate Effects (CCW), October 10, 1980, 1342 UNTS 137, reprinted in 19 ILM 1523 (1980), retrieved at http://www.unog.ch/80256EDD006B8954/

(httpAssets)/40BDE99D98467348C12571DE0060141E/$file/CCW+text.pdf., October 29, 2010.
16 See *Convention on Certain Conventional Weapons (CCW) At a Glance*, retrieved at http://www.armscontrol.org/factsheets/CCW., October 29, 2010.
17 CCW, note 5.
18 This argument would of course be contrary to the contentions of some robotics experts that lethal autonomous robots are particularly *unlikely* "to cause superfluous injury or unnecessary suffering."
19 Convention on the Prohibition of the Use, Stockpiling, Production and Transfer of Anti-Personnel Mines and on Their Destruction, September 18, 1997, 36 ILM 1507 (1997).
20 Convention on the Prohibition of the Development, Production and Stockpiling of Bacteriological (Biological) and Toxin Weapons and on Their Destruction (BWC), April 10, 1972, 26 UST 583, 1015 UNTS 163, retrieved at http://www.unog.ch/80256EDD006B8954/(httpAssets)/C4048678A93B6934 C1257188004848D0/$file/BWC-text-English.pdf, October 25, 2010.
21 Convention on the Prohibition of the Development, Production, Stockpiling and Use of Chemical Weapons and on Their Destruction (CWC), January 13, 1993, 1974 UNTS 45, retrieved at http://www.opcw.org/chemical-weapons-convention/articles/, October 25, 2010.
22 Treaty on the Non-Proliferation of Nuclear Weapons (NPT), opened for signature July 1, 1968, 21 U.S.T. 483, 729 U.N.T.S. 161 (entered into force March 5, 1970), retrieved at http://www.iaea.org/Publications/Documents/Infcircs/Others/infcirc140.pdf, October 25, 2010.
23 Ibid. at Art. IX.
24 Ibid. at Art. VI.
25 Ibid. at Art. II.
26 Organization of American States, Inter-American Convention on Transparency in Conventional Weapons Acquisitions, June 7, 1999, retrieved at http://www.oas.org/juridico/english/treaties/a-64.html, October 25, 2010.
27 Ibid.
28 CWC, Arts. I–II.
29 BWC, Art. I.
30 Comprehensive Nuclear Test Ban Treaty, opened for signature September 24, 1996, 35 I.L.M. 1439 (1996), retrieved at http://www.ctbto.org/the-treaty/treaty-text/, October 20, 2010.
31 Treaty Banning Nuclear Weapon Tests in the Atmosphere, in Outer Space and Under Water, August 5, 1963, 480 U.N.T.S. 43, 14 UST 1313, retrieved at http://disarmament.un.org/treatystatus.nsf/44e6eeabc9436b788525687700 78d9c0/35ea6a019d9e058a852568770079dd94?OpenDocument, October 20, 2010.
32 Ibid.

33 Treaty on Strategic Offensive Reductions (SORT), May 24, 2002, U.S.-Russ., 41 ILM 799 (2002), retrieved at http://moscow.usembassy.gov/joint_05242002.html, October 25, 2010. See also *The Strategic Offensive Reductions Treaty (SORT) At a Glance*, Arms Control Association, retrieved at http://www.armscontrol.org/factsheets/sort-glance, October 20, 2010.
34 Treaty on Conventional Armed Forces in Europe (CFE), November 19, 1990, 30 I.L.M., retrieved at http://www.dod.gov/acq/acic/treaties/cfe/index.htm, October 29, 2010.
35 NPT, Art. I.
36 CWC, Art. I.
37 Ibid.
38 BWC, Art. III.
39 Inter-American Convention on Transparency in Conventional Weapons Acquisitions.
40 International Court of Justice, *Legality of the Threat or Use of Nuclear Weapons*, Advisory Opinion, 1996 I.C.J. (July 8).
41 Rome Statute of the International Criminal Court, July 17, 1998, 2187 UNTS 3 (1998).
42 CWC.
43 BWC.
44 Protocol IV on Blinding Laser Weapons, annexed to Convention on Prohibitions or Restrictions on the Use of Certain Conventional Weapons Which May be Deemed to be Excessively Injurious or to Have Indiscriminate Effects, October 13, 1995, 35 I.L.M. 1218 (1996).
45 Ibid.
46 Ibid.
47 Jack Goldsmith, Eric A. Posner, *The Limits of International Law* (Oxford: Oxford University Press, 2005), 225.
48 Harald Hongjy Koh "Why Do Nations Obey International Law?" *Yale Law Review* 106 (1997), 2599.
49 Goldsmith, Posner, *The Limits of International Law*, 225.
50 Gross, *Moral Dilemmas of Modern War*, 238.
51 Alex Roland, "Keep the Bomb," 67–69. There is some evidence that since these rules were put in place in the second half of the 20th century—and since weaponry has become increasingly more lethal—warfare has killed fewer people, a decrease of some 82 percent compared to the first half of the century.
52 Robert D. Kaplan notes that:

> [T]he turn of the twenty-first century found the United States with bases and base rights in fifty-nine countries and overseas territories, with troops on deployments from Greenland to Nigeria to Singapore... Even before the terrorist attacks on the World Trade Center and the Pentagon on September

11, 2001, the U.S. Army's Special Operations Command was conducting operations in 170 countries per year.
Robert D. Kaplan, *Imperial Grunts, The American Military on the Ground* (New York: Random House, 2005), 7.

53 David H. Petraeus, "Letter to Soldiers in Iraq," March 15, 2007, retrieved at http://www.weeklystandard.com/weblogs/TWSFP/2007/03/petraeus_letter_to_the_troops asp, November 30, 2009.

54 FMI 3-07-22 Counterinsurgency Operations, retrieved at http://www.fas.org/irp/doddir/army/fm13-07-22.pdf, October 30, 2009.

55 John McCain in Pierre Atlas, "Even If It Works, US Shouldn't Torture,' *Real Clear Politics* (April 23, 2009), 2, retrieved at http://www.realclearpolitics.com/articles/2009/04/23/even_if_it_works_us_shouldnt_tortur..., November 30, 2009.

56 Fukuyama, *Our Posthuman Future*, 17.

57 Fukuyama, 194.

58 Ferguson, et. al. *Remapping Global Politics*, 342.

7
Conclusion

Abstract: *The innovation and adaption of technology is nothing new. It is the experience of humankind as she/he has proceeded through the millennia, often with great success and sometimes with considerable disaster. There appears to be something new afoot; technology is available democratically, it is innovated in a space of technological uncertainty and its power to change the way humans operate on all levels is staggering. Legal and ethical constraints on the military are of particular importance inasmuch as they constrain, at least in part, the unhampered use of force. Yet the governance of military technologies, while often different, cannot be separate from the governance of technologies in the 21st century generally.*

O'Meara, Richard Michael. *Governing Military Technologies in the 21st Century*. New York: Palgrave Macmillan, 2014. DOI: 10.1057/9781137449177.0011.

It appears clear that despite its myriad intricacies, technology is nothing new. Indeed, it forms the basis of culture itself, that description of how humankind operates when it forms into social organizations. On the one hand, technology constitutes merely the applied use of scientific knowledge to accomplish goals which occur to humans as they live their lives, competing with nature and other humans to survive, reproduce, and better their condition. It is then, as Bain opines, the "...tools, machines, utensils, weapons, instruments, housing, clothing, communicating and transporting devises and the skills by which we produce and use them."[1] On the other hand it is a good deal more, as Steigler proposes "...the pursuit of life by means other than life."[2]

Mankind has a good deal of experience with technology and the most perceptive among us are aware of its game changing abilities. Attempts have been made in some cultures, like the Chinese and the Japanese, to regulate certain technologies, especially when used for military purposes. Closing down commercial maritime industries, banning the use of gunpowder, and outlawing specific classes of weapons like the crossbow are but three examples. They all reflect attempts at the political and cultural level to direct resources and energies in directions that discredit or at least deemphasize military competitiveness. They also appear to be attempts, in part, to maintain the status quo. Most of these projects, however, appear to have failed, either because the failure to keep up has caused disastrous results when civilizations confronted others who had the advantage of emerging military technologies or because the central governance was unable to stifle grass roots innovation. Innovation, adaption, and especially use of technologies carry with them the prospect of change; change of competitive status, change in quality of life, or simply change which is interesting and attractive. Indeed, change would appear to be inevitable. As McNeil concludes, diffusion matters. People embrace change when someone from outside the community brings new things to their attention. It may be frightening or merely interesting, but it has the power to convince those in the community with the power to organize change, that it should be adopted.[3]

Where governance is fragmented as in an increasingly globalized environment, technology appears to thrive as well. Whether it be the military competitiveness of polities, the exuberance of individual accomplishment or simply the inquisitiveness of the human mind, where multiple spaces exist for the project of innovation, new technology emerges. Further, technology has traditionally diffused fairly rapidly from one civilization to another, each adapting it to its own needs and environments. The diffusion of gunpowder from Asia to the West is only one example. The characteristics of globalization, rapid communication, and movement of innovators between civilizations, only increase this diffusion.

Another characteristic of technology is that it carries with it both intended and unanticipated consequences. While innovation is important, most technological change occurs in a free-floating environment where adaption is practiced as a matter of course and unregulated diffusion occurs. Here, the regulators of culture, political, spiritual, economic, and military, cannot know the results of the technology. As J.N. Mathis points out, "[W]e will find ourselves caught off guard by changes in the political, economic, technological, strategic, and operational environments... Our goal is not to eliminate surprise—that is impossible. Our goal is, by a careful consideration of the future, to suggest the attributes of a joint force capable of adjusting with minimum difficulty when the surprise inevitably comes."[4] This is especially true in what Scranton refers to as an *environment of technological uncertainty*. Contemporary military competitiveness requires not merely more, quantitative, weapons but also better and different, qualitative, weapons. The innovation-to-use cycle has begun to move extremely rapidly with no time built in to examine the ramifications of the use, nor is there time to consider the legal, ethical, and moral appropriateness of the use.

Given, the above, it would appear that mankind has demonstrated the ability to accommodate change over the centuries, albeit with disastrous results for many. Civil society has introduced codes of ethics to regulate the conduct of innovators and adaptors, religious entities have promulgated practices and procedures regarding the uses to which technology should be put, and political organizations, both national and international, have entered into governance projects which recognize the worst uses of technologies and, in various forms, restrained those uses. IHL, for example, demonstrates one attempt to regulate the use of violence on the battlefield and the warrior code is a time honored attempt to reign in the worst practices which result from the unrestrained conduct of the strong over the weak.

Yet there seems to be more afoot today than merely the introduction of *the next big thing*. Human enhancement through the use of biotechnologies, nanotechnologies, robotics, and cyber technologies has the ability to turn on their head the assumptions upon which traditional restraints are based. While there is a good deal of discussion and disagreement regarding the exact nature and consequences of these changes, there can be little doubt that they are real, and represent existential challenges to the political, economic, social, philosophical, and military restraints with which humankind has become comfortable. Perhaps, more important, as discussed above, these changes are occurring in a climate of technological uncertainty and their ramifications—unanticipated consequences—threaten the actual survival of humankind. Concerns exist regarding the enhancement of the human species to the point where it is unrecognizable. Robotists inform us that they will have the ability within the very near future to fill the battle space with autonomous robots capable of lethality; neurobiologists envision a wide array of enhancements through pharmaceuticals and prosthetics which erase the physical and mental parameters which presently define human conduct on the battlefield; and cyber technologists, with their ability to permeate and disrupt every aspect of human life, are redefining the nature of warfare. For the civilian, these technologies represent challenges to the already difficult questions regarding the meaning of life, the distribution of resources, and the nature of *humanness*. Governance of these technologies in the 21st century may well be the most important project of humankind.

For the military, the human agency most likely to project violence with the use of these technologies, the stakes are equally high. Restraint must be made within the framework of a number of conflicting tensions: the responsibility to look to the safety of soldiers, the responsibility to insure competitiveness on the battlefield, and the responsibility to insure that the military is capable of carrying out its various missions as required by the state.

A number of models are available to the military policy-maker for the creation of restraints regarding emerging military technologies. These include military ethics, traditional IHL, and previous attempts to restrain the innovation, adaption, proliferation and use of weapons through international treaty regimes. There is also the possibility of creating new international treaties and practices, amending old ones, and forging new ethics for the use of new weapons.

At the end of the day, however, the military discussion is a subset—albeit an extremely important subset—of the discussion which must occur at the national and international level regarding these technologies. It appears that little has been done in this regard to date. This book argues that failure to act will not stop the use of these technologies. Rather, military technologies will continue to emerge with or without restraint, their unanticipated consequences are a matter of record. The *genie* is out of the bottle and his supervision is possible but not inevitable.

Notes

1 Bain, "Technology and State Government," 860.
2 Steigler, *Technics and Time*, 17.
3 McNeill, A History of the Human Community, xiii.
4 J.N. Mathis, *The JOE 2010*.

Bibliography

Adams, Thomas K. "Future Warfare and the Decline of Human Decisionmaking," *Parameters* (Winter 2001), 57–71.

Agrere, Sam. *Good Governance, Promoting Good Governance: Principles, Practices and Perspectives.* London: Commonwealth Secretariat, Marlboro House, 2000.

Al-Sharif, Sayyid Imam. *Rationalizing Jihad in Egypt and the World, 2007* as cited in Jared Brachman, "Al Qaeda's Dissident," *Foreign Policy, Special Edition,* 2009.

Allhoff, Fritz, Patrick Lin, James Moor, John Weskert, eds. *Nanoethics, the Ethical and Social Implications of Nanotechnology.* Hoboken, New Jersey: John Wiley & Sons, Inc., 2007.

Allhoff, Fritz and Patrick Lin, "What's so Special about Nanotechnology and Nanoethics?" *International Journal of Applied Philosophy,* 2 (2): 179–90, 2006.

Altman, Jurgen. *Military Nanotechnology, Potential Applications and Preventive Arms Control.* New York: Routledge, 2006.

Annas, George. "'The Man on the Moon, Immortality and Other Millennial Myths: The Prospects and Perils of Human Genetic Engineering," *Emory Law Journal* 49: 3 (Summer 2000).

Annas, George, Michael A. Grodin. *The Nazi Doctors and the Nuremberg Code: Human Rights in Human Experimentation.* New York: Oxford University Press, 1992.

Anton, Captain. "A Short History of the English Longbow." *Archers of Ravenwood,* retrieved at http://

www.archers.org/default.asp?section=History&page=longbow, November 23, 2009.

Arkin, Ronald. *Governing Lethal Behavior in Autonomous Robots*, Boca Raton, FL: CRC Press, 2009.

——. "Governing Lethal Behavior: Embedding Ethics in a Hybrid Deliberative/Hybrid Robot Architecture," Report GIT-GVU-07-11, Atlanta GA: Georgia Institute of Technology's GVU Center: http://www.cc.gatech.edu/ai/robot-lab/online-publications/formalizationv35.pdf.

Arthur, W. Brian. *The Nature of Technology, What It Is and How It Evolves.* New York: The Free Press, 2009.

Asimov, Isaac. *I, Robot.* New York: Bantam Books, 1950.

Atlas, Pierre. "Even If It Works, US Shouldn't Torture," *Real Clear Politics*, April 23, 2009 retrieved at http://www.realclearpolitics.com/articles/2009/04/23/even_if_it_works_us_shouldn't_torture, November 30, 2009.

Bekey, George. *Autonomous Robots: From Biological Inspiration to Implementation and Control*, Cambridge, MA: MIT Press, 2005.

Bernstein, Steven, Jennifer Clapp, Mathew Hoffmann. "Reframing Global Environmental Governance, Results of a CIGI, CIS Collaboration." *The Center for International Governance Innovation*, Working Paper no. 45, December 2009.

Bonvillian, William. "POWER PLAY, THE DARPA Model and the U.S. Energy Policy." *Holidays*, November/December, 2006.

Boot, Max. "Are We the Mongols of the Information Age?" *Los Angeles Times* Op-ed, October 29, 2006.

——. *War Made New: Weapons, Warriors and the Making of the Modern World.* New York: Penguin (USA), 2006.

——. "The Paradox of Military Technology," *The New Atlantis: Journal of Technology & Society*, Fall, 2006.

Bozeman, Barry, Catherine Slade, and Paul Hirsh. "Understanding Bureaucracy in Health Science Ethics: Toward a Better Institutional Review Board," *American Journal of Public Health* 99, no. 9 (Summer 2009).

Breuil, Henri, Raymond Lautier, trans. B.B. Rafter. *The Men of the Old Stone Age*, Westport, Ct.: Greenwood Press, 1980.

Brouma, Ian. *Inventing Japan.* New York: Modern Library, 2004.

Bull, Stephen. *Encyclopedia of Military Technology and Innovation.* Portland, OR: Greenwood Pub, 2004.

Canning, John, G.W. Riggs, Holland, O. Thomas, Carolyn Blakelock, 2004. "A Concept for the Operation of Armed Autonomous Systems on the Battlefield", Proceedings of Association for Unmanned Vehicle Systems International's (AUVSI) Unmanned Systems North America, August 3–5, 2004, Anaheim, CA.

Canning, John. 2008. "Weaponizing Unmanned Systems: A Transformational Warfighting Opportunity, Government Roles in Making it Happen", Proceedings of Engineering the Total Ship (ETS), September 23–25, 2008, Falls Church, VA.

Cardwell, Donald. *The Fontana History of Technology*. London: Fontana Press, 1994.

Carroll, Patrick. *Science, Culture, and State Formation*. Berkley, CA: University of California Press, 2006.

Chambers, John ed. *The Oxford Companion to American Military History*. Oxford: Oxford University Press, 1999.

Chameau, Jean-Lou, William F. Ballhaus, and Herbert S. Lin eds. *Emerging and Readily Available Technologies and National Security: A Framework for Addressing Ethical, Legal, and Societal Issues*, Washington, DC: The National Academies Press, 2014.

Chapell, Paul K. *Will War Ever End? A Soldier's Vision of Peace for the 21st Century*. Rvive Books, 2009.

Clements, Patrick J. "Research and Development in the FY 2010 Defense Budget." *Budget Insight,* Stimson Center Blog, November 3, 2009, retrieved at http://budgetinsight.wordpress.com/2009/11/03/research-and-development-in-the-fy-2010-defense-budget/, November 20, 2009.

Coleman, Kevin. "Technology Driven National Security Strategy." *Directions Magazine*, Feb. 2004 retrieved at http://www.directionsmag.com/printer.php?articleid=521, December 02, 2009.

Contamine, Philippe, ed. *War and Competition between States*. Oxford: Oxford University Press, 2000.

Copeland, Robert M. ed. "The SIrUS Project, Towards a determination of which weapons cause 'superfluous or unnecessary suffering.'" International Committee of the Red Cross: Geneva, 1997 *retrieved at* http://www.icrc.org, January 17, 2010.

Cozzens, Susan. "Emerging Technologies and Inequalities: Beyond the Technological Transition." Draft Comments, Technology, Policy and Assessment Center, School of Public Policy, Georgia Institute of Technology, April 5, 2009.

Crowley, Roger. "The Guns of Constantinople," *HistoryNet.com* retrieved at http://www/historynet.com/theguns-of-constantinople.htm, November 24, 2009.

Davidson, Neil. *"Non-Lethal" Weapons*. New York: Palgrave Macmillan, 2009.

Department of Defense, *FY2009–2034 Unmanned Systems Integrated Roadmap*. 2009.

——. Directive No. 3000.3, *Policy for Non-Lethal Weapons*, July 9, 1996.

——. "Defense Advanced Research Projects Agency Strategy Plan, May 2009".

Diamond, Jared. *Guns, Germs and Steel, The Fates of Human Societies*. New York: W.W. Norton & Co, 1999.

——. *Collapse, How Societies Choose to Fail or Succeed*. Penguin Books: New York: W.W. Norton, 2006.

Dado, Malcolm, ed. *Non-Lethal Weapons: Technological and Operational Prospects, Jane's* on-line Special Report. November, 2000.

Drutman, Lee. "Where Does Innovation Come from?" *Science & Environment*, September 2009, *retrieved at* http://www.miller-mccune.com/scienceenvironment/where-does-innovation-come-from, November 23, 2009.

Dunlap, Charles J. "Technology and the 21st Century Battlefield: Recomplicating Moral Life for the Statesman and the Soldier." Carlisle Barracks, PA: Strategic Studies Institute, January 15, 1999.

Duke, C. and K. Dill. "The Next Technological Revolution: Will the US Lead or Fall Behind?" retrieved at http://www.biophysics.org/pubaffairs/revolution.pdf, November 25, 2009.

Durant, Will and Ariel. *The Lessons of History*. New York: Simon & Shuster, 1968.

Duffy, Christopher. *The Military Experience in the Age of Reason*. New York: Barnes & Noble, 1997.

——. *Siege Warfare: The Fortress in the Early Modern World, 1494–1660*. Reissued New York: Routledge, 1996.

Edge, David, and John M. Paddock. *Arms and Armor of the Medieval Knight*. New York: Crescent Books, 1998.

Exhart, William G. "Lawyering for Uncle Sam When He Draws His Sword," retrieved at www.law.umkc.cau/faculty/projects/trials...exhart.html, November 23, 2009.

Fainaru, Steve. 2008. *Big Boy Rules: America's Mercenaries Fighting in Iraq*. Philadelphia, PA: Da Capo Press.

———. "Soldier of Misfortune." *The Washington Post* (Monday, December 1, 2008): C1–C2.

Federer, William J. ed. *A Treasure of Presidential Quotations*. Amerisearch: Ashtabula: OH, 2004.

Ferguson, Yale H., Richard W. Mansbach. *Remapping Global Politics, History's Revenge and Future Shock*. Cambridge: Cambridge University Press, 2004.

Fink, Sheri, "U.S. Medical Personnel and Interrogations: What Do We Know? What Don't We Know?" *ProPublica*, April 9, 2009.

French, Shannon, *And the Code of the Warrior: Exploring Warrior Values Past and Present*. Lanham, Md: Rowman & Littlefield Pub. Inc.

Frese, Pamela R., Margaret C. Harrell, eds. *Anthropology and the United States Military, Coming of Age in the Twenty-first Century*. New York: Palgrave Macmillan, 2003.

Fukuyama, Francis. *Our Posthuman Future, Consequences of the Biotechnology Revolution*. New York: Picador, 2002.

———. "How to Regulate Science," *The Public Interest*, no. 146, 2002, 3–22.

Giles, Keir with Andrew Monaghan, *Legality in Cyberspace: An Adversary View*, Carlisle, PA: Strategic Studies Institute, U.S. Army War College, March 2014.

Glenn, David. "Anthropology Association Formally Disapproves of Military Program." *The Chronicle of Higher Education*, November 7, 2007, retrieved at http://chronicle.com/article/Anthropology-Association/29909, January 15, 2010.

Goldsmith, Jack. *The Terror Presidency, Law and Judgment inside the Bush Administration*. New York: W.W. Norton, 2007.

Gray, Colin. *Another Bloody Century: Future War*. London: Phoenix Paperbacks, 2006.

Goliath, "Informed Consent in the Military: Fighting a Losing Battle against the Anthrax Vaccine." *American Journal of Law & Medicine*, June 22, 2002, retrieved at http://goliath.ecnext.com/coms2/gi_0199-1833062/informed-consent-in-the-military.html, November 25, 2009.

Gross, Michael L. *Bioethics and Armed Conflict, Moral Dilemmas of Medicine and War*. Cambridge, MA: MIT Press, 2006.

———. *Moral Dilemmas of Modern War, Torture, Assassination, and Blackmail in an Age of Asymmetric Conflict*. Cambridge: Cambridge University Press, 2010.

———. "Medicalized WEAPONS & Modern WAR," *Hastings Center Report* 40, no 1 (2010): 34–43.

Grossman, Dave. "Evolution of Weaponry, A Brief Survey of Weapons Evolution: The Roman System." *Killogy Research Group*, 1999, retrieved at http://www.killogy.com/art_weap_sum_roman.htm, November 23, 2009.

Guetlein, Mike. "Lethal Autonomous Weapons-Ethical and Doctrinal Implications." Naval War College Joint Military Operations Paper, February 2005.

Guston, David H. John Parsi, Justin Tosi. "Anticipating the Ethical and Political Challenges of Human Nanotechnologies," in Fritz Allhoff, Patrick Lin, James Moor, John Weckert, *Nanoethics, the Ethical and Social Implications of Nanotechnology.* Hoboken, NJ: Wiley & Sons, 2007.

Hacker, Barton C., American *Military Technology, the Life Story of a Technology.* Baltimore: The Johns Hopkins University Press, 2006.

Habermas, Jurgen. *1984–1987. The Theory of Communicative Action,* vols 1/11, trans. Thomas McCarthy. Boston, MA: Beacon Press.

——. "Bestiality and Humanity: A War on the Border between Law and Morality." *Kosovo: Contending Voices on Balkan Interventions,* ed. William Joseph Buckley. Grand Rapids, MI: William B. Eerdmans, 2000.

——. *The Postnational Constellation.* Cambridge, MA: MIT Press, 2001.

——. *Der gespaltene Westen [The Divided West].* Frankfurt: Suhrkamp Verlag. (Trans. Jeffrey Craig Miller. Cambridge: Polity Press, 2006).

Hanson, Victor. *The Western Way of War, Infantry Battle in Classical Greece,* 2nd edn. Berkley: University of California Press, 1989.

Halberstam, David. *War in a Time of Peace: Bush, Clinton and the Generals.* New York: Simon & Schuster, 2001.

Harris, P.M.G. *The History of Human Populations: Migrations, Urbanization and Structural Change,* v. 11. Westport, CN: Praeger Publishing, 2003.

Hawking, Stephen, "Transcendence Looks at the Implications of Artificial Intelligence—But Are We Taking AI Seriously Enough?" *The Independent* (May 4, 2014) retrieved at http://www.independent.couk/news/science/stephen-hawking-transcendence-looks-at-theimplications-of-artificial-intelligence, May 4, 2014.

Hirschland, Mathew J. "Information warfare and the New Challenges to Waging Just War," (APSA Annual Meeting, San Francisco, California, March 7, 2001).

House Armed Service Committee, "Statement by Dr. Tony Tether, Director Defense Advanced Research Projects Agency, before the Subcommittee on Terrorism, Unconventional Threats, and Capabilities House Armed Services Committee, United States House of Representatives," March 27, 2007.

Howard, Michael, Peter Paret ed. and trans. *Carl von Clausewitz: On War.* Princeton, NJ: Princeton Univ. Press, 1976.

———. *War in Human History.* Oxford: Oxford University Press, 1976.

Hooker, Richard. *Enlightenment Glossary, World Civilizations,* 1996, retrieved at http://www.wsu.edu:8080/-dee/GLOSSARY/MODERN.HTM, November 24, 2009.

Huntington, Samuel P. *The Soldier and the State: The Theory and Politics of Civil-Military Relations.* Cambridge, MA: The Belknap Press of Harvard University Press, 1957.

Institute on Medicine as a Profession, *ETHICS ABANDONED, Medical Professionalism and Detainee Abuse in the War on Terror,* Columbia University: 2013.

International Committee of the Red Cross, "International Humanitarian Law-treaties and Documents, retrieved at http://www.icrc.org/ihl.nsf/TOPICS?OpenView, November 03, 2009.

Jason, the Mitre Corporation. *Human Performance,* JSR-07-625, Study performed on behalf of the Office of Defense Research and Engineering, Project no. 13079022, March 2008.

Joy, Bill. "Why the Future Doesn't Need Us" in Fritz Allhoff, Patrick Lin, James Moor, John Weckert, eds. *Nanoethics: The Ethical and Social Implications of Nanotechnology.* Hoboken, NJ: Wiley & Sons, 2007.

Kahn, David. *Seizing the Enigma: The Race to Break the German U-Boats Codes, 1939–1943.* New York: Barnes & Noble, 2009.

Kaplan, Robert D. *Imperial Grunts: The American Military on the Ground.* New York: Random House, 2005.

Keegan, John. *A History of Warfare.* New York: Knopf, 1993.

———. *The Price of Admiralty, the Evolution of Naval Warfare.* New York: Penguin Group (USA), 1990.

———. *The Face of Battle.* New York: Penguin Group (USA), 1978.

Keeley, Lawrence. *War before Civilization: The Myth of the Peaceful Savage,* New York: Oxford University Press, 1996.

Keiper, A. "The Nanotechnology Revolution." *The New Atlantis,* Summer (2) 19, 2003.

Kennedy, David. "The International Rights Movement: Part of the Problem?" *Harvard Human Rights Journal,* v. 15, Spring, 2002.

Klein, John J. "The Problematic Nexus: Where Unmanned Combat Air Vehicles and the Law of Armed Conflict Meet," *Air & Space Power Journal-Chronicles on Line Journal,* July 22, 2003, retrieved at http://www.airpower.maxwell.af.mil/airchronicles/cc/klein.html, November 20, 2009.

Kramer, Franklin D., Stuart H. Starr, Larry K. Wentz, eds. *Cyberpower and National Security.* Washington, DC: Potomac Books, Inc., 2009.

Krishnan, Armin. *Killer Robots, Legality and Ethicality of Autonomous Weapons.* Burlington, Vt.: Ashgate Publishing Co., 2009.

Koplow, David A. *Non-Lethal Weapons: The Law and Policy of Revolutionary Technologies For the Military and Law Enforcement.* Cambridge: Cambridge University Press, 2006.

Kuhlau, Frida. Stefan Erikson, Kathinka Evers, "Taking Due Care, Moral Obligations in Dual Use Research." *Bioethics,* v. 22, no. 9, November 2008.

Kurzweil, Ray. *The Singularity Is Near: When Humans Transcend Biology.* New York: Penguin Books, 2005.

——. "On the National Agenda: U.S. Congressional Testimony on the Societal Implications of Nanotechnology" in Fritz Allhoff, Patrick Lin, James Moor, John Weckert. *Nanoethics: The Ethical and Social Implications of Nanotechnology.* Hoboken, NJ: Wiley & Sons, 2007.

——. *The Age of Spiritual Machines: When Computers Exceed Human Intelligence.* New York: Penguin Group, 2000.

——. *The Age of Intelligent Machines.* Boston: MIT Press, 1999.

Kuttab, Daoud. "Indiscriminate Attack," in *Crimes of War, the Book, What the Public Should Know.* New York: W.W. Norton & Co., 2007.

Lamaadar, Alia. "War and Peace from Weapons Technology: Examining the Validity of Optimistic/Semi-Optimistic Technological Determinism." *The McGill Journal of Political Studies,* 2003–4.

Landers, John. "The Destructiveness of Pre-Industrial Warfare: Political and Technological Determinants." *Journal of Peace Research,* v. 42, no. 4, 2005, 455–70.

Landler, Mark, John Markoff, Steven Lee, "Digital Fears emerge After Data Siege in Estonia" (*New York Times*: May 24, 2007), 2, retrieved at http://www.nytimes.com/2007/05/29/technology/29estonia.html?_r=1, October 20, 2010.

Langhorne, Richard. *The Coming of Globalization: Its Evolution and Contemporary Consequences.* New York: Palgrave Macmillan, 2001.

Lazarski, Anthony J., "Legal Implications of Uninhabited Combat Aerial," *Air & Space Power Journal-Chronicles Online Journal*, (March 27, 2001), retrieved at http://www.airpower.maxwell.af.mil/airchronicles/cc/lazarski.html, November 17, 2009.

Lederer, Susan E. "Chpt 17. The Cold War and Beyond: Covert and Deceptive American Medical Experimentation," in *Military Medical Ethics*, v. 2. Washington, D.C.: Borden Institute, Office of the Surgeon General, 2003.

Lele, Ajey. "Technologies and National Security." *Indian Defense Review*, v. 24, no. 1, Jan–Mar, 2009.

Levine, Eugenia, "Command Responsibility, The Mens Rea Requirement," *Global Policy Forum* (February 2005) retrieved at http://www.globalpolicy.org/component/content/article/163/28306.html, December 10, 2009.

Lewand, Kathleen. "A Guide to the Legal Review of New Weapons, Means and Methods of Warfare, Measures to Implement Article 36 of Additional Protocol I of 1977," International Committee of the Red Cross, revised November 2006.

Levanthes, Louise. *When China Ruled the Seas: The Treasure Fleet of the Dragon Throne, 1405–1433*. Oxford: Oxford University Press, 1994.

Lewer, Nick. "Non-lethal Weapons." *Forum for Applied Research and Public Policy*, v. 14, no. 2, Summer 1999.

Lewis, Bernard. *What Went Wrong? Western Impact and Middle East Response*. Oxford: Oxford University Press, 2002.

Lifton, Robert Jay. *The Nazi Doctors: Medical Killing and the Psychology of Genocide*. New York: Basic Books, 1986.

Lin, Patrick, George Bekey, Keith Abney, *Robots in War: Issues of Risk and Ethics*. California Polytechnic State University. Paper sponsored by the U.S. Department of the Navy, Office of Naval Research under Award #N00014-1-1152.

Lin, Patrick, Fritz Allhoff, "Nanoscience and Nanoethics: Defining the Disciplines." in Fritz Allhoff, Patrick Lin, James Moor, John Weckert eds. *Nanoethics: The Ethical and Social Implications of Nanotechnology*. Hoboken, NJ: Wiley & Sons, 2007.

Lin, Patrick, Kieth Abney, George B. Bekey, eds. *Robot Ethics: The Ethical and Social Implications of Robotics*, Cambridge, Ma: The MIT Press: 2012.

Livi-Bacci, Massimo. *A Concise History of World Population*, 4th edn. Oxford: Blackwell Publishing, 2007.

Lucas, George R. Jr. 2001. *Perspectives on Humanitarian Military Intervention.* Response by General Anthony C. Zinni, U.S. Marine Corps (retired). "The Fleet Admiral Chester W. Nimitz Memorial Lecture Series on Security Affairs-University of California at Berkley." Berkley, CA: University of California Institute of Government Studies, 2001.

———. "The Role of the International Community in the Just War Tradition: Confronting the Challenges of Humanitarian Intervention and Preemptive War," *Journal of Military Ethics* 2, no.2, 2003.

———. "From *Jus ad bellum* to *Jus ad pacem:* Rethinking Just War Criteria for the Use of Military Force for Humanitarian Ends." *Ethics and Foreign Intervention.* eds. Donald Scheid and Deen K. Chatterjee. New York: Cambridge University Press, 2004.

———. "Defense or Offense: Two Streams of Just War Tradition." *War and Border Crossings: Ethics when Cultures Clash.* eds. Peter A. French and Jason A. Short. Lanham, MD: Rowman & Littlefield, 2005.

———. "Methodological Anarchy: Arguing about War, and Getting it Right," *Journal of Military Ethics,* 6, no. 3: 246–52, 2007.

———. "Inconvenient Truths: Moral Challenges to Combat Leadership in the new Millennium," 20th Annual Joseph Reich, Sr. Memorial Lecture, U.S. Air Force Academy (November 2007). http://www.usna.edu/Ethics/Publications/Inconvenient%20Truths%20USAF%20Academy.pdf.

———. "*Jus in bello* for Nonconventional Wars." San Francisco, CA: International Studies 49th Annual Meeting (March 26), retrieved at http://www.allacademic.com/meta/p_mla_apa_research_citation/2/5/3/9/p253594_index.html

———. *Anthropologists in Arms: the Ethics of Military Anthropology.* Lanham, MD: AltaMira Press, 2009.

———. "'This Not Your Father's War'—Confronting the Moral Challenges of 'Unconventional' War." *Journal of National Security Law & Policy,* v. 3, no. 2: 329–40, 2009.

———. "'New Rules for New Wars': Jus in Bello for Irregular War. 'Theory and Methodology of Ethical Conduct during Asymmetric War.'" New Orleans, LA: International Studies Association 51st Annual Meeting (February 11, 2010).

———. "'Forgetful Warriors.' Neglected Lessons on Military Leadership from Plato's *Republic.*" *The Ashgate Research Companion to Modern*

Warfare. eds. George Kassimeris and John Buckley. London: Ashgate Press, 2010.

Lucas, George R., Rubel, 2007: *Ethics and the Military Profession:Tthe Moral Foundations of Leadership*, revised edition. New York: Longman/Pearson Publishers, 2007.

Mann, Charles C. *1491: New Revelations of the Americas before Columbus*. New York: Vintage Books, 2006.

Marturano, Antonio. "When Speed Truly Matters, Openness Is the Answer." *Bioethics* v. 23, no.7, September 2009.

Mayer, Alexander. "Analysis: A Look at US Airstrikes in Pakistan through September 2009." *The Long War Journal*, October 1, 2009, retrieved at http://www.longwarjournal.org/archives/2009/10/analysis_us_airstrik.php, November 25, 2009.

McManus, John, Sumeru G. Mehta, Arnette R. McClinton, Robert A. DeLorenzo, Tony W. Baskin. "Informed Consent and Ethical Issues in Military Medical Research," *U.S. Army Academy of Emergency Medicine*, v. 12, no. 11, November 2005.

McNeill, William H. *The Pursuit of Power, Technologies, Armed Force, and Society since A.D. 1000*. Chicago: The University of Chicago Press, 1982.

——. *The Rise of the West*. Chicago: University of Chicago Press, 1992.

——. *A History of the Human Community*. Englewood Cliffs, NJ: Prentiss-Hall, Inc., 1988.

McNaughler, Thomas L., "The Real Meaning of Military transformation: Rethinking the Revolution," *Foreign Affairs*, v. 86, no. 1 (January/February, 2007), 140–7.

Melo-Martin, Immaculafda de. "Chimeras and Human Dignity," *Kennedy Institute of Ethics Journal*, v. 18, no. 4, December 2008.

Miles, Stephen. "Oath Betrayed: Torture, Medical Complicity and the War on Terror," 2006.

——. "Military Medicine and Human Rights." *The Lancet*, v. 364, no. 9448, November 20, 2004.

Minsky, Martin. *Semantic Information Processing*. Cambridge, MA: MIT Press, 1968.

Mirillo, Stephen. "Guns and Government, A Comparative Study of Japan and Europe," *Journal of World History*, v. 6, no. 1, 1995.

Misa, Thomas J., Philip Brey, Andrew Feenber eds. *The Compelling Tangle of Modernity and Technology*. Cambridge, MA: MIT Press, 2004.

Moore, Daniel, "Nanotechnology and the Military," in Fritz Allhoff. Patrick Lin, James Moor, John Weckert eds., *Nanoethics: The Ethical and Social Implications of Nanotechnology*. Hoboken, NJ: Wiley & Sons, 2007.

Montreaux, 2008. "Montreaux Document on Pertinent International Legal Obligations and Good Practices for States Related to Operations of Private Military and Security Companies during Armed Conflict." Montreaux, Switzerland: International Committee of the Red Cross (September 17, 2008), retrieved at http://www.icrc.org/web/eng/siteengo.nsf/htmlall/montreuxdocument-170908/$FILE/Montreux-Document.pdf, November 27, 2009.

Moreno, Jonathan D. *Mind Wars, Brain Research and National Defense*. New York: Dana Press, 2006.

Neuneck, Gotz and Christian Alwart, "The Revolution in Military Affairs, Its Driving Forces, Elements and Complexity." Interdisciplinary Research Group; on Disarmament, Arms Control and Risk Technologies, Working Paper 5 #12 May 12, 2008.

Office of the Secretary of Defense. "Annual Report to Congress, Military Power of the People's Republic of China, 2009".

O'Meara, R.M (2011) in conjunction with G. Marchant, B. Allenby, R. Arkin, E. Barrett, J. Borenstein, L. Gaudet, O. Kittrie, P. Lin, G. Lucas, J. Silberman, *International Governance of Autonomous Military Robots,* in *Columbia Science and Technology Law Review*.

Orend, Brian. *The Morality of War*. Peterborough, Ontario: Broadview Press, 2006.

O'Roarke, Ronald. "Unmanned Vehicles for U.S. Naval Forces: Background and Issues for Congress," CRS Report for Congress, updated April 12, 2007.

Pacey, Arnold. *Technology in World Civilization*. Cambridge, MA: MIT Press, 1990.

Parker, G. *The Military Revolution, Military Innovation and the Rise of the West, 1500–1800,* 2nd edn. Cambridge: Cambridge University Press, 1996.

Petraeus, David. "Letter to Soldiers in Iraq," March 15, 2007, retrieved at http://www.weeklystandard.com/weblogs/TWSP/2007/03/petraeus_letter_to_the_troops.asp, November 30, 2009.

——. *Counterinsurgency,* eds. David H. Petraeus and James F. Amos. Army Field Manual 3–24. Washington, DC: U.S. Government Printing Office.

Perrin, Noel. *Giving Up the Gun, Japan's Reversion to the Sword, 1543–1879.* Boston: D.R. Godine, 1988.

Pfaff, Tony, *Resolving Ethical Challenges in an Era of Persistent Conflict*, U.S. Army War College: 2011.

Physicians for Human Rights. White Paper, "Broken Laws, Broken Lives," November 28, 2007.

Pielke, Robert A. Jr. *The Honest Broker: Making Sense of Security Policy and Politics.* Cambridge: Cambridge University Press, 2000.

President's Council of Advisors on Science and Technology. "National Nanotechnology Initiative: Second Assessment and Recommendations of the NNAP," April 2008.

President's Council on Bioethics. "Beyond Therapy: Biotechnology and the Pursuit of Happiness, October, 2003," retrieved at http://bioethicsprint.bioethics.gov/reports/beyondtherapy/chapter=1.html, November 10, 2009.

Ponsford, Mathew "Robot Exoskeletons Suits That Could Make Us Superhuman," CNN, May 22, 2013, retrieved at http://cpf.cleanprint.net/cf/cpf?action=print&type=filePrint=cnn&url=http%3A%2F%2Fwww.cnn.com, May 4, 2014.

Ratner, Daniel, Mark Ratner. *Nanotechnology and Homeland Security: New Weapons for New Wars.* Upper Saddle River, NJ: Pearson Education, Inc., 2004.

Reid, William, *Arms Through the Ages.* New York: Harper Collins, 1976.

Roland, Alex. "Presentation Notes at the Teaching the History of Innovation Workshop," published in *Footnotes, Foreign Policy Research Institute,* retrieved at http://fpri.org/footnotes/1402.200902.roalnd.wartechnology.html, November 22, 2009.

——. "Keep the Bomb," *Technology Review*, August/September 1995.

——. "Technology and War," *American Diplomacy,* 1997, retrieved at http://www.unc.edu/depts/diplomat/D_ISSUES/AMDIPL_4/ROLAND.HTML, November 24, 2009.

Rosen, Stephen P. "New Ways of War: Understanding Military Innovation," *International Security,* Summer, 1988.

Rosenau, James N. *Along the Domestic-Foreign Frontier: Exploring Governance in a Turbulent World.* Cambridge: Cambridge University Press, 1997.

Rosenberg, Harold L. *Atomic Soldiers: American Victims of Nuclear Experiments.* Boston: Beacon Press, 1980.

Rout, Mathew William. "To Define & Control: The Utility of Military Ethics in the New Zealand Army's Contemporary Operations

Environment." Master's Thesis, University of Canterbury, 2009, retrieved at http://cantebur6y.ac.nz/handle/10092/3048, November 26, 2009.

Rowe, Neil C. "Ethics of Cyber War Attacks", in Lech J. Janczewski and Andrew M. Colarik, eds. *Cyber Warfare and Cyber Terrorism*, Hershey, PA: Information Science Reference, 2008.

———. "The Ethics of Cyberweapons in Warfare," *International Journal of Cyberethics*, v. 1, no. 1, 2009.

———. "War Crimes from Cyberweapons," *Journal of Information Warfare*, v. 6, no. 3, 2007, 15–25.

Runciman, Steve. *The Fall of Constantinople, 1453*. Cambridge: Cambridge University Press, 1965.

Standler, Ronald B. "Response of Law to New Technology," May 8, 1997, retrieved at http://www.rbs2.com/lt.htm, November 25, 2009.

Schachtman, Noah. "How Technology Almost Lost the War in Iraq, The Critical Networks are Social-Not Electronic." *Wired Magazine*, no. 15.12, November, 27, 2007.

Showalter, David. *Railroads and Rifles, Soldiers, Technology and the Unification of Germany*. Hamden, CN: Archon Books, 1975.

Scranton, Philip. "The Challenge of Technological Uncertainty," *Technology and Culture*, v. 50, no. 2, April 2009, retrieved at http://muse.jhu.edu.proxy.libraries.rutgers.edu/journals.technology_and_culture/v050/50, November 20, 2009.

Sharkey, Noel. 2007 "Robot Wars are a Reality", *The Guardian* (UK), August 18, 2007, retrieved at http://www.guardian.co.uk/commentisfree/2007/aug/18/comment.military, October 11, 2009.

———. "Automated Killers and the Computing Profession", *Computer* 40: 122–4, 2007.

———. "Cassandra or False Profit of Doom: AI and War", *IEEE Intelligent Systems,* Jul/August 2008, p. 1417.

———. "Grounds for Discrimination: Autonomous Robot Weapons", *RUSI Defense Systems*, 11: 2, 86–9, 2008.

Sklerov, Mathew J. "Solving the Dilemma of State Response to Cyberattacks: A Justification for the Use of Active Defenses against States Who Neglect Their Duty to Prevent." *Military Law Review*, v. 201, Fall, 2009.

Skolnikoff, Eurgen B. *The Elusive Transformation: Science, Technology and the Evolution of International Politics*. Princeton: Princeton University Press, 1993.

Singer, Charles. ed. *A History of Technology*. Oxford: Clarendon Press, 1954–84 vols 1–5.

Singer, Peter W. *Wired For War: The Robotics Revolution and Conflict in the 21st Century*, New York: Penguin Group, 2009.

———. "How to Be All That You Can Be: A Look at the Pentagon's Five Step Plan for Making *Iron Man* Real." Brookings Institution, November 17, 2009.

———. "Military Robots and the Law of War," *The Atlantis: Journal of Technology* and *Society*. Winter, 2009.

Sloan, John. "The Stirrup Controversy, posted on discussion list medieval@ukanvm.cc.ukans.edu on 5 October, 1994 as part of the thread 'The Stirrup Controversy,'" retrieved at http://www.fordham.edu/halsall/med/sloan.html, on November 23, 2009.

———. "Once More into the Stirrups: Lynn White Jr., Medieval Technology and Social Change." *Technology and Culture*, v. 44, no. 3, July 2003.

Spagnolo, Antonio G. "Outlining Ethical Issues in Nanotechnologies," *Bioethics* v. 23, no. 7, September 2009.

———. "Should We Enhance Animals?" *Journal of Medical Ethics*, v. 35, no. 11, November 2009.

Sparrow, Robert. "Building a Better WarBot: Ethical Issues in the Design of Unmanned Systems for Military Applications." Original Paper, June 23, 2008.

———. "Killer Robots." *Journal of Applied Philosophy*, v. 24, no. 1, 2007.

———. "Predators or Plowshares? Arms Control of Robotic Weapons." *IEEE Technology and Society Magazine*, Spring 2009.

Stoker, Liam, "Military Exoskeletons Uncovered: Ironman Suits a Concrete Possibility," *Army Technology Market & Customer Insight* (January 30, 2012) retrieved at http://ww.army-technology.com/featuremilitary-exoskeletons-uncoered-ironman-suits-a-concrete-possibility, April 29, 2014.

Tikk, Eneken, Kadri Kaska, Kristel Ruuimeri, Mari Kert, Anna-Maria Taliharm, Liis Vihul. *Cyber Attacks Against Georgia: Legal Lessons Identified*. Tallinn, Estonia: Cooperative Cyber Defense Centre of Excellence, 2008.

Tuchman, Barbra. *The Proud Tower: A Portrait of the World before the War 1890–1914*. New York: The Macmillan Company, 1966.

United Nations Economic and Social Commission for Asia and the Pacific, "What Is Good Governance?" United Nations, 2009, retrieved at http://www.unescap.org/pdd/prs/ProjectActivities/Ongoing/gg/governance.asp, November 20, 2009.

U.S. Army Field Manual 3-07-22, Counterinsurgency Operations retrieved at http://www.fas.org/irp/doddir/army/fm13-07-22.pdf, October 30, 2009.

U.S. Army Field Manual 27–10, The Law of Land Warfare.

United States Senate, Republican Policy Committee, "The Perils of Universal Jurisdiction," December 18, 2006.

Van Creveld, Martin. *Technology and War, From 2000 B.C. to the Present*, rev. edn New York: The Free Press, 2004.

Verchio, Donna Marie. "Just Say No! The SIrUS Project: Well-intentioned, but Unnecessary and Superfluous." *The Air Force Law Review*, Spring, 2001, v. 51.

Waltz, Kenneth N. *Man, the State and War: A Theoretical Analysis*. New York: Colombia University Press, 2001.

Wallach, Wendall, Colin Allen, *Moral Machines: Teaching Robots Right From Wrong*. Oxford: Oxford Univ. Press, 2009.

Waltzer, Michael. *Just and Unjust Wars*. New York: Basic Books, 1977.

Wesley, Russell F. *The American Way of War: A History of United States Strategy and Policy*. Bloomington, IN: The Indiana University Press. 1973.

Welsh, Cheryl. "2006 Government Mind Control Debate," *Mind Justice*, 2006.

Williams, Trevor. *The History of Invention*. New York: Facts on File, 1987.

Williamson, Murray, "War and the West," *Orbis*, Philadelphia, PA, no. 2 (Spring, 2008), 350.

Wilson, Peter. "Revolutions in Military Affairs as Ways of War, 1914–2014." Presentation Remarks at the Strategic Implications of Emerging Technologies Conference, XX Strategy Conference, U.S. Army War College, Carlisle Barracks, Pa, April 2009.

White, Lynn Jr. *Medieval Technology and Social Change*. Oxford: Oxford University Press, 1996.

Wolfendale, Jessica, Steve Clarke, "Paternalism, Consent, and the Use of Experimental Drugs in the Military." *Journal of Medicine & Philosophy*, v. 33, no. 4, August 2008.

Womack, "The Development and Recent Applications of the Doctrine of Command Responsibility: With Particular Reference to the *Men's Rea* Requirement," in Sienho, Yee, ed. *International Crime and Punishment,* Selected Issues. Lanham, MD: University Press of America, 2003.

Young, Simon, *Designer Evolution: A Transhumanist Manifesto.* Amherst, NY: Prometheus Books, 2007.

Index

American Medical Association (AMA) 69
American Psychiatric Association 69
American Psychological Association (APA) 69, 71
American Society of Civil Engineers 69
Anthrax 56
Artificial Intelligence 15, 17
Autonomous Weapons (AW) 15

Bilibid prison 53
bionanobots 12
bioscience 14

Chinese Way of War 32–33
cognitive therapies 3
Comprehensive Nuclear Test Ban treaty (CTBT) 87
Conventional Armed Forces in Europe Treaty 88
Convention on Prohibitions or Restrictions on the Use of Certain Conventional Weapons (CCW) 85
Convention on the Prohibition of the Development, Production and Stockpiling of Bacteriological (Biological) and Toxin Weapons and on Their destruction (BWC) 86
Convention on the Prohibition of the Development, Production, Stockpiling and Use of Chemical Weapons and on the Destruction (CWC) 86
Convention on the Prohibition of the Use, Stockpiling, Production and transfer of Anti-Personnel Mines and on Their Destruction 86
cyber technology 19–21

Defense Advanced Research Projects Agency (DARPA) 13, 36, 37–38, 72
Department of Defense, FY2009–2034 Unmanned Systems Integrated Roadmap 15–16
dual-use technology 37

Eisenhower, Dwight D. 36
exoskeleton 17

FM 3-07-22 Counterinsurgency Operations 92

good governance 67, 68

Hippocratic Oath 68
human enhancement 12–14

Human Terrain System Program (HTS) 57–58
Human Universal Load Carrier (HULC) 18
The Hurt Locker 15

Improvised Explosive Devices (IEDs) 15, 58
industrial revolutions 3, 4
information technology 5, 59
informed consent 54–56
innovation 25–26
Institutional Review Board (IRB) 54
Inter-American Convention on Transparency in Conventional Weapons Acquisitions 86
International Committee for Robot Arms Control (ICRAC) 71–76
International Committee of the Red Cross 73
International Humanitarian Law (IHL) 52, 72–75, 90–9
I, Robot 15
Ironman 69

Joy, Bill 7, 8
Judge Advocates
  the conscious of the command 83, 84
Jus ad bellum 47
Jus in bello 47
Jus post bellum 47

Law of Armed Conflict (LOAC) 67

matrix 15
military ethics 47–50
military function 47
military-industrial complex 36
modernity 31–32

nanoscience 11–12
nanotechnologies 7, 11–12

National Security Presidential Directive (NSPD) 54/Homeland Security Presidential Directive 23
NBIC (Nao, Bio, Info, and Cognitive technologies) 11
Network of Concerned Athropogists (NCA) 57–58
non-lethal weapons 18–19
  Acoustic Rays 19
  Chemical Calmatives or Malodorants 19
  Directed Energy Heat Rays 19
  pepper spray 19
  Projectile Netting 19
  Taser 19
Nuremberg Principles/Nuremberg Code 53–54

Physicians for Human Rights 70
President's Council of Advisors on Science and Technology, 2008 71–72
President's Council on Bioethics, 2003 14, 71
prohibition on certain weapons 72–73
  command responsibility 74
  discrimination 74
  lawfare 75
  military necessity 73
  proportionality 73
  universal jurisdiction 75
Protocol 1V on Blinding Laser Weapons, annexed to Convention on Prohibitions or Restrictions on the Use of Certain Conventional Weapons Which May be Deemed to be Excessively Injurious or to Have Indiscriminate Effects 85

Remote Environmental Monitoring Unit (REMUS) 17
Revolutions in Military Affairs (RMAs) 4–6

Robotic Evacuation Vehicle (REV) 17
Robotic Extraction Vehicle (REX) 17
Robotics 14–18

scientia 45
SIrUS criteria 72–73
Star Trek 83–84
stirrup 28–29

technocracies and technopolies 45
technological uncertainty 25, 34–36
The Terminator 15
Transcendence 15
Treaty Banning Nuclear Weapon Tests in the atmosphere, in Outer Space and Under Water 87
Treaty on Strategic Offensive Reductions (SORT) 87

Treaty on the Non-Proliferation of Nuclear weapons (NPT) 86
Tuskegee Syphilis Study 53

unanticipated consequences 48, 51, 61
United States Army values 47
United States Code of Conduct 46
United States Oath of Office 46
unmanned aircraft systems (UASs) 17
unmanned surface vehicles (UGVs) 17
unmanned surface vehicles (USVs) 7
unmanned underwater vehicles (UUVs) 17

warrior-builder-diplomat spirit 92–93
warrior culture 48–50

The manufacturer's authorised representative in the EU is Springer Nature Customer Service Centre GmbH, Europaplatz 3, 69115 Heidelberg, Germany. If you have any concerns regarding our products, please contact ProductSafety@springernature.com

Printed and bound by CPI Group (UK) Ltd, Croydon, CR0 4YY

23/03/2026

02076402-0017